ROBOTICS - LEGAL, ETHICAL AND SOCIOECONOMIC IMPACTS

Edited by **George Dekoulis**

Robotics - Legal, Ethical and Socioeconomic Impacts
http://dx.doi.org/10.5772/66612
Edited by George Dekoulis

Contributors

Johan Hoorn, Robert Dupre, Vasileios Argyriou, Roderick Beard, Domenico Parisi, Claudio Germak, Luca Giuliano, Maria Luce Lupetti, Sara Khan, Joel Perez, Jose Carlos Moreno, Alejandro Zornoza, Jose Luis Guzman, Francisco Rodriguez, Julian Sanchez-Hermosilla, Vicente Matellan, Camino Fernández-Llamas, Ángel Manuel Guerrero-Higueras, Francisco Javier Rodríguez-Lera, George Dekoulis

Published by InTech

Janeza Trdine 9, 51000 Rijeka, Croatia

Publishing Process Manager Romina Skomersic
Technical Editor SPi Global
Cover InTech Design team

First published December, 2017
Printed in Croatia
Legal deposit, Croatia: National and University Library in Zagreb

Additional hard copies can be obtained from orders@intechopen.com

Robotics - Legal, Ethical and Socioeconomic Impacts, Edited by George Dekoulis
p. cm.
Print ISBN 978-953-51-3635-4
Online ISBN 978-953-51-3636-1

For EU product safety concerns:
IN TECH d.o.o., Prolaz Marije Krucifikse Kozulić 3, 51000 Rijeka, Croatia,
info@intechopen.com or visit our website at intechopen.com.

Meet the editor

Prof. George Dekoulis received a PhD degree in Space Engineering and Communications from Lancaster University, UK, in 2007. He was awarded a first-class BEng (Hons) in Communications Engineering from De Montfort University, UK, in 2001. He received several awards from STFC, UK, and EPSRC, UK, and the 'IET Hudswell International Research Scholarship'. He is currently a professor in Aerospace at Aerospace Engineering Institute (AEI), Cyprus. He was previously a professor in Electrical and Electronics Engineering at Middle East Technical University, Turkey. He has previously worked as a professor in Aerospace Engineering at various departments, such as Aeronautical and Space Engineering, Professional Flight, Robotics, Mechatronics, Computer Engineering, Electrical and Electronics and Mechanical Engineering. His research is focused on the design of reconfigurable aerospace engineering systems.

Contents

Preface

This edited volume is a collection of reviewed and relevant research chapters, concerning the developments within the *Robotics: Legal, Ethical and Socioeconomic Impacts* field of study.

The book includes scholarly contributions by various authors and is edited by a group of experts pertinent to roboethics. Each contribution comes as a separate chapter complete in itself but directly related to the book's topics and objectives.

The book is divided into six sections: 'Introduction', 'Robotics', 'Legal Impacts', 'Ethical Impacts', 'Socioeconomic Impacts' and 'Epilogue'.

After 'Introduction', the first section, 'Robotics', includes chapters dealing with the topics of risk estimation in robotics and the impact of human behaviour and trajectory tracking error using fractional-order PID control law for two-link robot manipulator via fractional adaptive neural networks.

The following section, 'Legal Impacts', includes chapters on 'Robot's Liability: A Use Case and a Potential Solution' and 'Cybersecurity of Robotics and Autonomous Systems: Privacy and Safety'.

The section 'Ethical Impacts' includes chapter titled 'Ethic Reflections about Service Robotics from Human Protection to Enhancement: Case Study on Cultural Heritage'.

'Socioeconomic Impacts' section presents two chapters: 'Mechanical Empathy Seems Too Risky (Will Policymakers Transcend Inertia and Choose for Robot Care?): The World Needs It' and 'Electronic Prescribing and Robotic Dispensing: The Impact of Integrating Together on Practice and Professionalism'.

'Human, Not Humanoid, Robots' is included in the last section of this book, 'Epilogue'.

The target audience comprises scholars and specialists in the field.

InTechOpen

Section 1

Introduction

Introductory Chapter: Introduction to Roboethics - The Legal, Ethical and Social Impacts of Robotics

George Dekoulis

Additional information is available at the end of the chapter

http://dx.doi.org/10.5772/intechopen.71170

There is no difference between a deep learning humanoid robot and the finest of humans. They are simply our reflection.

1. Introduction

Would you have entered this Boeing 737 that just lands? The robotic co-pilot Aircrew Labour In-cockpit Automation System (ALIAS) has been used in simulation mode to fully fly and land a Boeing 737, as shown in **Figure 1**. The project was developed by the Defense Advanced Research Projects Agency (DARPA).

All of us have entered an airplane before. An airplane is a robot mimicking a bird. In the same way, a humanoid robot mimics a human.

As the available scientific processing power is constantly increasing, we will constantly be learning about or using advanced products with large memory modules and artificial intelligence (AI) capabilities. Therefore, it is highly possible in the future to also enter a semi-autonomous, as in **Figure 1**, or a fully-autonomous airplane too.

The usage of highly intelligent products that continue learning, while we are using them, exhibit both positive and negative features. Every manufacturer considers its products as being totally positive. Through usage it is the users that usually realise the drawbacks of an AI product. This represents the social impact aspect of using AI robotic products. Therefore, the need for introducing a new higher-level field of science would evaluate the virtues of AI robotic products as early as possible in the design process. This recently developed field of science is called Robot Ethics or Roboethics in its shortened version. Currently, the field of AI robotics is broadly considered to be unlegislated. However, regulatory bodies, such as the EU, are trying to develop legal directives in order to gradually proceed to the full legislation of AI robotic products.

Figure 1. Robotic co-pilot ALIAS landing a Boeing 737.

The first step in solving a problem is to understand the problem. In order to understand the magnitude and significance of the problem, bear in mind that currently is impossible even for the manufacturers to estimate the current state of consciousness and knowledge of a deep learning AI robot that a customer operates.

2. Aim of the book and organisation

The legal, ethical and social aspects of using AI robots represent the aim of this book. The collective effort of distinguished international researchers has been incorporated into one textbook suitable for the broader audience interested into the scientific field of Roboethics.

Chapters 1 and 2 represent the results of recent scientific work on Robotics.

- Chapter 1 presents the emerging topic of automated risk assessment in a domestic scene. It focuses on safer human and robotic interactions within a given environment. Hazards are identified and the risk factor is quantified.
- Chapter 2 presents a solution to the problem of measuring the trajectory tracking error using fractional-order PID control law for two-link robot manipulator. The results exhibit a satisfactory performance of the fractional-order dynamical neural network with online learning.

Chapters 3 and 4 report the significant progress achieved in the field of legislating the operation of AI Robotic systems.

- Chapter 3 presents a system of distribution of responsibility of damages caused by robots. Legal and ethical aspects of robotics are presented. The European Parliament adopted the Report with recommendations to the Commission on Civil Law Rules on Robotics (2015/2103(INL)) on February 2017. The liability level ranges according to the robot's learning capability and the knowledge learned from its owner. A responsibility setting matrix is proposed by the authors.
- Chapter 4 discusses the emerging topic of cyber-security of Robotics and autonomous systems in terms of privacy and safety. A survey on cyber-security attacks associated to service robots is presented. A taxonomy is presented that classifies the risks faced by the users when using robots. A robot software development phase is finally presented for preventing unauthorised access to service robots.
- Chapter 5 discusses some of the ethical impacts related to the usage of service robots. A case study on cultural heritage is presented.

Chapters 6 and 7 analyse major socioeconomic impacts that arise from the usage of AI robotic systems.

- Chapter 6 analyses the social impacts that arise through the usage of healthcare robots. Emphasis is given to the implementation of care robots as a direct solution to the Global Ageing problem. Immediate governmental support is required in terms of both allocating funds for R&D and regulating the legal frame regarding care robots.
- Chapter 7 analyses the positive impacts to the healthcare system that can be achieved, if reliable electronic prescribing and robotic dispensing techniques are applied into hospital pharmacies. The efficiency, safety, professionalism and reliability of the hospital pharmacies would greatly be increased. More patient focused activities can therefore be developed.

Chapter 8 is the Epilogue of this book: looking into the mirror

- Chapter 8 presents an interesting, philosophical and controversial discussion to the topics covered so far. It is looking at the topics covered so far through the mirror. We have put so much effort in developing humanoid robots, i.e. AI robots that resemble and act like humans. Why not simulate and replicate human behaviour through computer simulations. If the results match human behaviour, we would understand ourselves and the construction of human communities. Such simulated results of human robots and societies are presented. Some of the problems that human robots will pose to human beings are discussed.

3. Conclusion

We hope this book will increase the sensitivity of all the community members involved with Roboethics. The significance of incorporating all aspects of Roboethics right at the beginning of the creation of a new deep learning AI robot is emphasised and analysed throughout the book. AI robotic systems offer an unprecedented set of virtues to the society. However, the principles of Roboethical design and operation of deep learning AI robots must be strictly legislated, the manufacturers should apply the laws and the knowledge development of the AI robots should be closely monitored after sales. This will minimise the drawbacks of implementing such intelligent technological solutions. These devices are a representation of ourselves and form communities like us. Learning from them is a way to improve ourselves.

Author details

George Dekoulis

Address all correspondence to: dekoulis@aeispace.org

Aeronautical and Space Engineering Department, Aerospace Engineering Institute (AEI), Nicosia, Cyprus

Section 2

Robotics

Chapter 2

Risk Estimation in Robotics and the Impact of Human Behaviour

Rob Dupre and Vasileios Argyriou

Additional information is available at the end of the chapter

http://dx.doi.org/10.5772/intechopen.69770

Abstract

Within this chapter the emerging topic of automated risk assessment in a domestic scene is discussed, state of the art techniques are reviewed followed by developed methodologies which focus on safer human and robotic interactions with an environment. By using the risk estimation framework, the notion of a quantitative risk score is presented. Hazards within a scene are evaluated and measured using risk elements which provide a numeric representation of specific types of risk. Emphasis is given to the concept of risk as a result of interaction with an environment, specifically whether human or robotic actions in a scene can effect overall risk. To this end, techniques which simulate human or robotic behaviour with regard to risk in an environment are reviewed. Specifically the ideas of interaction and visibility are addressed defining risk in terms of areas within a scene that are visited most often and which are least visible. As with any behaviour simulation techniques, validation of their accuracy is required and a number of simulation evaluation techniques are reviewed. Finally a conclusion as to the current state of automated risk assessment is given, with a brief look at the future of the research area.

Keywords: risk estimation, domestic robotics, risk perception for robots, scene analysis, pedestrian simulation, simulation analysis

1. Introduction

The concept of automated risk analysis for domestic scenes is a developing area of research, where the goal is to allow the quantification of measurable elements of risk into some sort of coherent risk score. With the definition of a risk score, thresholds can be defined to trigger certain actions, this might be a cue for a robot to find an alternative route through an environment or to take action to reduce the discovered risk. Humans and robots are almost constantly exposed to risks in one form or another. As humans, risks are identified and appropriate actions are taken almost on a subconscious level and as a result we are excellent at self preservation. To date robots have lacked these mechanisms despite an equal need to ensure their own protection.

By providing the functionality to detect these potential risks, advantages are gained in two ways, firstly the robots themselves are given an awareness of potentially dangerous scenarios and therefore become better able to protect themselves from damage. Secondly, the risk detection abilities can be utilised for those at risk members of society whose hazard perception capabilities are diminished or underdeveloped.

As an example, most people will make use of a kitchen in their daily routine and the act of leaving a knife near the edge of the sideboard would not pose a significant risk (**Figure 1**). Consider now the same scenario but with an elderly adult suffering from Parkinson's disease or early onset dementia being the user of the kitchen, or with a child running around the room. The possibility of knocking that knife on the floor for a healthy adult is low, but for a child not actively aware of their surroundings or those suffering from a disability this could become more of a likelihood, likewise without suitable software governing interactions, a robot could equally suffer from the results of a falling blade.

Invariably there is a percentage of society that could be classed as more *at risk* than the rest. For example there are an estimated 10 million disabled people living in the UK [2] and the elderly (65+years) and young children (< 10years) account for 28% of the UK population [3]. People within this age range are statistically more likely to have an accident in the home [4]. Additionally the number of people that fall into the elderly category is increasing. Europe has one of largest ageing populations with 24% already aged 60years or over, this is set to increase sharply such that by 2050 that proportion is projected to reach 34% [5]. As such the infrastructure which provides continued support services to this population will come under increasing pressure. With this increase in the amount of elderly adults continuing to live self sufficiently, the need to alleviate the stresses on the support services and the need to provide a high level of

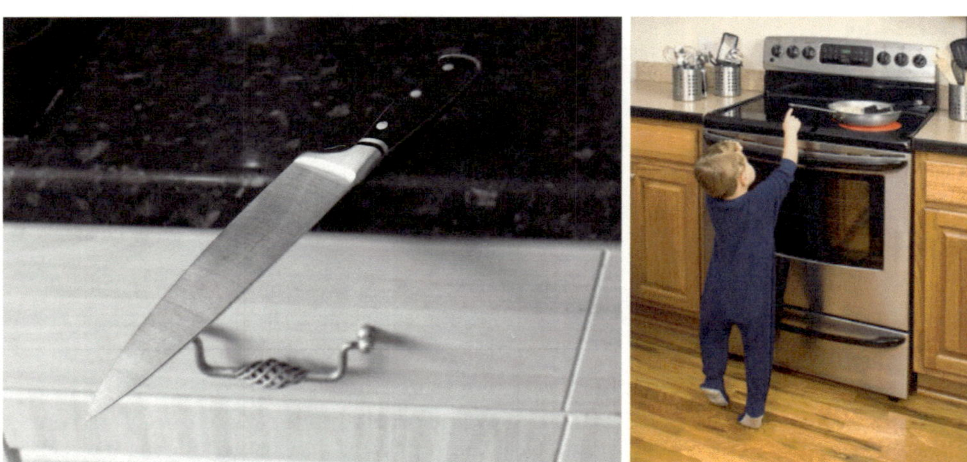

Figure 1. Example situations: A knife left precariously at the edge of a table, a unsupervised child and a hot cooking pan [1].

care, emerging technologies can be used to ensure they remain safe and if something were to happen, appropriate services can be notified or actions taken.

The International Federation of Robotics (IFR) have noted a steady increase in the sales of professional and personal robotics since 2012, with predictions that during the period of 2014–2107 the estimated number of service robots sold for domestic/personal use could be as high as 27million [6]. With this emergence of more and more domestic robotics, these devices also face similar risks and require similar systems by which to identify risk. This may be within the bounds of keeping an at risk user safe, for example a domestic robot ensuring that a room is safe for a child, or just keeping the robotic system itself operational by ensuring the path it intends to take is free from potential hazards [7].

As such automated risk assessment research provides the facility to measure risk, resulting in a numerical scaling upon which thresholds of risk can be defined and appropriate actions taken. The increase in the availability of these technologies and the integrated vision hardware that is often incorporated, provides the ability to analyse hazards for those at risk individuals within a domestic environment. The primary form of computing these risk scores is through the use of a risk estimation framework which provides a basis from which a combination of measurable elements of risk can be output into a risk score for a given environment. Existing approaches to the definition and measurement of hazards include the detection of risk related objects [8, 9] scene stability [10, 11] and human behaviour simulation [12, 13].

One of the primary issues when looking at a unified approach to risk estimation is the problem of context. In short the importance of ensuring that a provided risk score is relevant to the end user regardless of situation. This is due to the fact that what can be considered safe in one environment may not be in others. For example, a container of liquid at the edge of a table is risky in a household environment however in a chemical laboratory this might pose a far larger danger. In this case stability estimation would need to take precedent over the prevalence of hazard features or behaviour analysis. Similarly users of the environment will also affect how risk is perceived; if the environment contains children or elderly adults the threshold of what is considered risky may need to change. However, regardless of context, the elements that might contribute to the concept of risk can be broken down into components from which a decision can be made. These components include elements such as shape, size, material, temperature, position as well as many others.

An area of risk evaluation that has yet to be fully explored is the effect interaction with an environment has on a potential risk. Take again the example of a knife balanced at the edge of a table, alone this poses no hazard, however if there is a likelihood that there will be some interaction with the knife then the scenario develops into something more hazardous. Additionally the human response to risk opens up potential new threats as the behaviour of an agent in a scene will differ based on the presence of hazards. As such the ability to define these interactions and, on an environment basis, predict them allows for a more accurate assessment of risk in a scene.

To this end the focus of this chapter is the introduction of automated risk analysis research methods that can be used in scene analysis tasks to measure risk. As such the forthcoming sections will continue as follows; Firstly an overview of the base risk estimation framework [14] with a brief analysis of some existing risk elements. Next a focus is given to the concepts of risk analysis relating to interaction through the use of behaviour simulation. As such simulation techniques relating to risk are introduced which can be utilised as prediction mechanisms, evaluating likely interactions between environment and user. Finally simulation evaluation is discussed, ensuring that suitable and appropriate validation techniques are used to verify the simulation outputs.

2. Research methods

Within the following sections a overview of existing risk estimation techniques will be given. Starting with a base framework from which different measurable elements of risk can be combined to give a robust and relevant evaluation of a scene's risk.

2.1. The risk estimation framework

The risk estimation framework as defined in Ref. [14], outputs a risk score from measurable elements of risk. As such the cumulative risk score, R, for a scene is defined as the weighted sum of n measured risk elements e (Eq. (1)). The weighting specified for each element should fall into a range from zero to one, with the sum of the weightings for all included elements being equal to one.

$$R = \sum_{i=1}^{n}(w_i e_i) \tag{1}$$

The use of this weighting of elements allows the consideration of the aforementioned issues regarding the context in which the risk is found. Allowing the final system to tailor its outputs, applying more weighting to elements that are more relevant in a given situation.

2.1.1. Risk elements

A risk element represents any scene property that may present a danger and is given as a quantitative measure that could highlight those potential risks. An obvious example would be temperature obtained from a thermal camera. To date existing work is risk estimation has provided techniques, and thereby risk elements, for the detection of risk related objects [8, 9] and scene stability [10, 11]. With this concept of risk elements and an established, flexible framework, the concept of defining risk in terms of environment interaction and behaviour analysis allows for smart enabled homes or domestic robots to analyse an unknown setting and build up a detailed map of the potential risk.

2.2. Risk related behaviour simulation

In June 1883, the magazine The Chautauquan, posed the question, 'If a tree were to fall on an island where there were no human beings would there be any sound?'. The proposed answer

http://dx.doi.org/10.5772/intechopen.69770

was no, given that sound is a human's perception of movement in airwaves. If there is no one there to perceive those movements then there cannot be any sound. This concept is loosely the basis for risk analysis based on interaction. If there is a boiling beaker of acid, perched precariously close to the edge of a table, but no-one ever goes near it. Is it still a risk? In reality, yes, but far less so that if that table was situated in the middle of a busy train station at rush hour. As such a method is needed by which the prediction and analysis of these interactions can be used to contribute to the risk scores detected within a scene.

2.2.1. *Environmental risk maps*

To simulate interaction with an environment it is important to consider a number of aspects. Firstly and most notability, the expected paths that humans or robots would take through a scene (**Figure 2**). Second to this is the concept of visibility, how much of the scene is visible to the average human as they navigate the environment. For example if a path passes near a table, the top of that table is likely visible but below or behind would not be. Additionally the concept of redirection on account of a hazard must also be taken into consideration, allowing for the fact that paths though an environment may change on account of a discovered risk. As a result of this redirect, other areas within the environment would have an increased likelihood of interaction.

Figure 3 outlines this process. Initially a scene is preprocessed to obtain a 2D mapping or floor plan, interest points are then extracted and paths through the scene are simulated. The environmental risk map is then calculated, providing localised risk scores for the whole scene. Finally, simulations can be re-run based on the presence of risks detected by any other part of the risk estimation framework.

2.2.2. *Interaction and visibility maps*

By tracking a human or robot's movement through an environment over time a map can be created describing which areas are used more frequently than others. However method requires

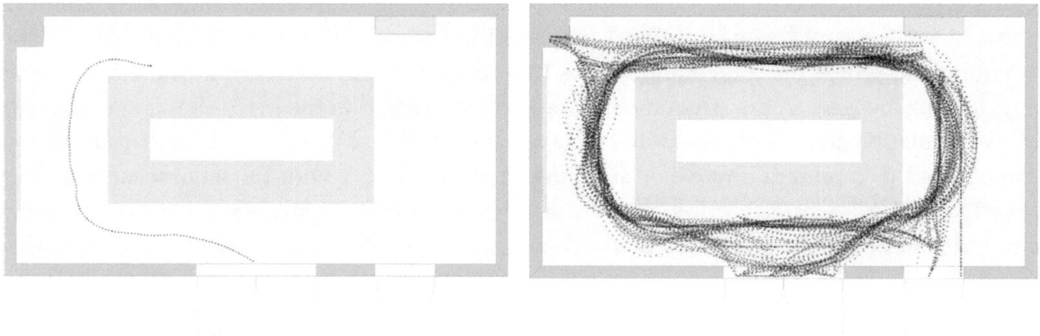

(a) (b)

Figure 2. Example scene with simulated paths overlaid. (a) A single path simulated through the room, (b) a compound image of all simulated paths to/from entrance and exits and to/from facilities within the room such as seating, computers and other likely destinations.

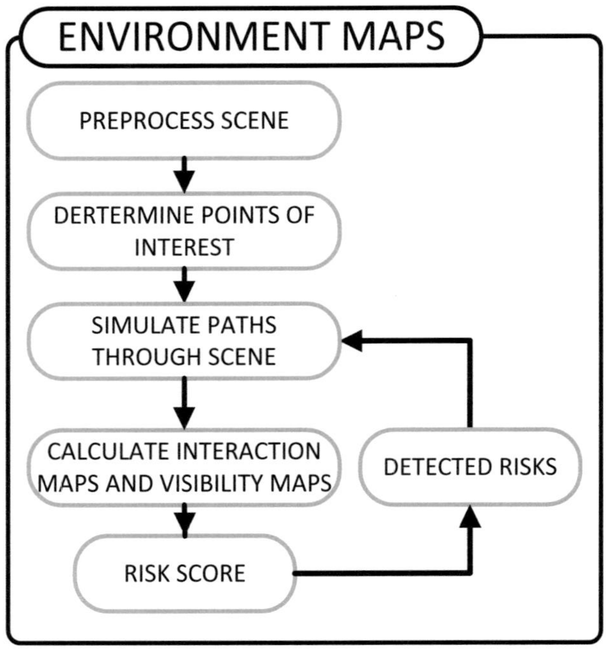

Figure 3. Flow diagram illustrating the stages required to calculate an environmental risk score for a scene.

long term observation and is specific to each individual environment. As such utilisation of behaviour simulation techniques allows for similar results in a fraction of the time.

Initially to allow simulation of an environment to be run, mapping of that environment is required. This can be achieved through a number of methods [15–18] with the detection of entrances through use of existing techniques [19]. By creating a low resolution two dimensional map of an environment, with labels defining points of interest, multiple simulations can be run to replace the long term monitoring techniques.

Using a two dimensional Cartesian representation of an environment a movement space in terms of x and y coordinates is defined. The mapping represents a low resolution view of the environment, where one unit square maps to a set square measurement in the environment (e.g. $0.5m$). Obstacles in the environment are also represented in the map, allowing the agents in the simulation an awareness of what is traversable and what is not. An agent in a simulation can be used to represent anyone or anything that can interact with the environment, in our case it could be a at risk adult/child or a domestic robot. **Figure 4a** shows an example environment in which a meeting room is shown with a set of tables arranged in the middle of the room, **Figure 4b** shows the 2D environment map of the same scene.

Using the defined entrance/exit locations and points of interest, an exhaustive set of paths that take into account all possible path connotations can be made. For each one of these paths a simulation is run in which an individual agent, representing a human or robotic subject, traverses the environment from a start location to a destination, avoiding any obstacles that may be in their way.

(a) (b)

Figure 4. Example environment. (a) Captured photo, (b) 2D mapping (orientated to match the photo), where differing shades represent (from light to dark): half height obstacles, entrance/exits or points of interest, traversable areas and finally full height obstacles/non-traversable area.

The results of an individual simulation provides positions $P = [x, y]^t \in \mathbb{R}^2$ for an agent at a specific time t. By removing the time component and plotting these points in the movement space we build up a picture of a single traversed path, similarly plotting all the paths from all the agents demonstrates areas that are most commonly used (**Figure 5**). 2D histograms are then created by binning each individual position, of each agent, at each time, into its respective unit measure within the environment map (Eq. (2)).

$$h_{i,j}(P) = \begin{cases} 1, & \text{iff } P_x \in i \text{ and } P_y \in j \\ 0, & \text{otherwise} \end{cases} \tag{2}$$

where i and j represent bins or unit measures within the scene along the x and y axis respectively of the environment map. The sum of points within each bin is used as a measure of frequency, (Eq. (3)). The result is a low resolution frequency map, indicating areas of high and low interaction (**Figure 6**).

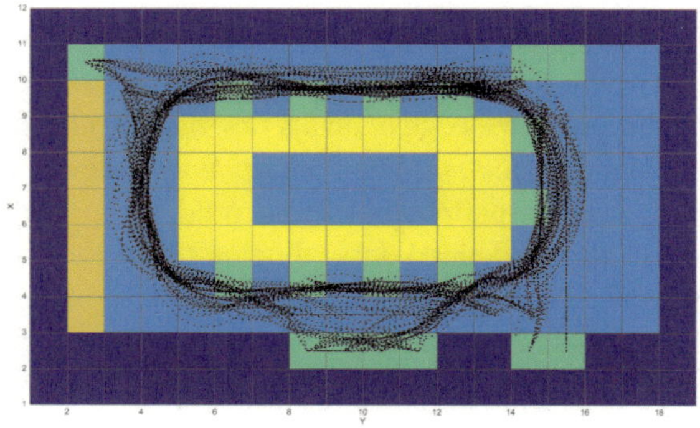

Figure 5. Simulated paths overlaid on environment map for all possible path connotations.

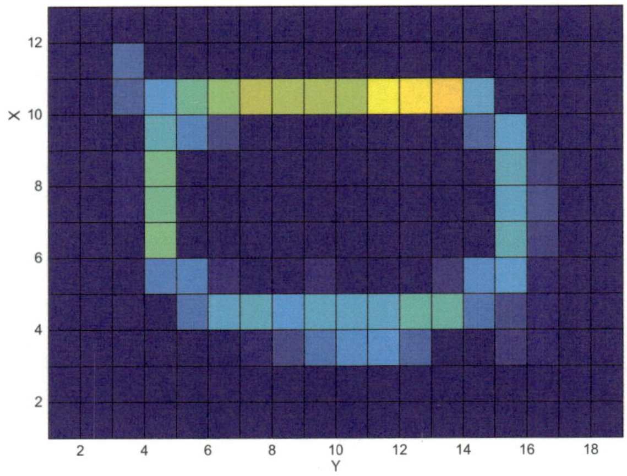

Figure 6. Resultant interaction map as a result of the simulated paths and the binning process.

$$\mathbf{H}_{i,j} = \sum_{k=1}^{t} h_{i,j}(P_k) \tag{3}$$

Thus locations of high frequency define areas in the environment in which the simulation algorithm estimates a higher level of human presence, as such these areas would present more of a risk than those of low frequency.

Using the same simulation techniques and histogram principle as Eqs. (2) and (3), the concept is expanded to encompass the visibility component of simulated agents in an environment. Each agent within the simulation has a number of defined properties, these include agent radius, movement speed, acceleration and turning speed. In addition to these a number of properties are defined that pertain to that agent's ability to *see* the environment. A field of view is defined as $\mathbf{F} = [\varphi, q] \in \mathbb{R}^2$ subject to v, which specifies the angular range of that agents peripheral vision φ, as well as a viewable radius q

$$v = f(\mathbf{F}_{\varphi,q}) \tag{4}$$

where v is a value from zero to one defining how well an agent can see at a specific point of their vision. In the case where the agent is representing a human, closer and more central points are better seen and those at the edge of an agent's vision and further away rated worse. This can be based on a logarithmic scale [20] for humans, however a more comprehensive model of visual acuity in human peripheral vision could be applied. Likewise for robotic vision, visual acuity would also be defined depending on the sensing devices equipped to the device. For example, the concept of diminishing acuity at the edge of perception may not be applicable. During the simulation an agent's viewable area is recorded per time instance, subject to an agent's properties. Within the context of the environment mapping this is defined as whether an agent can see a specific unit square of the map or not (**Figure 7**).

Parameters for field of view and peripheral visual acuity allows the tailoring of the environmental risk maps to better reflect those that use the environment. As in the cases of children,

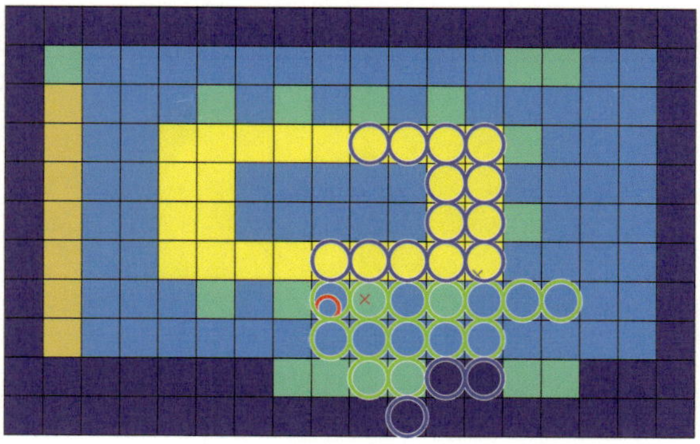

Figure 7. Example frame of a simulation with the agent's current field of view overlaid. Light circles indicate seen traversable area, dark circles indicate seen obstacles. The smaller circle is the current position of the agent and the cross is the direction of movement.

the visually impaired, or the elderly who have reduced peripheral visual acuity or Tunnel Vision due to conditions such as glaucoma or brain damage the risk of visibility maybe a more important factor, likewise for robots vision maybe less of a consideration than just the interaction component.

The 2D map of the scene is updated to reflect the differing heights of obstacles, such that a label is defined for obstacles that can be seen over and for obstacles that cannot (**Figure 4b**). For example, walls fully obscure the agents view, however low height obstacles, such as tables, block vision directly behind them but allow vision further away. As can be seen in **Figure 7**, where the area in the centre of the room cannot be seen as it is occluded by the presence of the tables, where as the table across is still visible to the agent.

As before simulations are run for the given connotations of paths. Position is extended such that $P = [x, y, v]^t \in \mathbb{R}^3$ where v represents how well that position was *seen* by the agent at that time subject to (Eq. (4)), as they navigate through the scene on their estimated path (**Figure 7**). The 2D histogram continues to bin based on location, however the contribution to the bin is now made by the visibility component (Eq. (5)).

$$h_{i,j}(P) = \begin{cases} P_v, & \text{iff } P_x \in i \text{ and } P_y \in j \\ 0, & \text{otherwise} \end{cases} \tag{5}$$

The visibility map is then given as the summation of the histogram bins over time as (Eq. (3)).

Figure 8a gives an example of a single agent's track with the compounded visibility map for that path shown in **Figure 8b**. The same method is used for all the given path combinations and a single compound visibility map returned for that environment (**Figure 9**).

The higher the visibility values recorded for an individual unit square of the environment map, the more often and better seen that area of the environment is likely to be. Conversely those areas with low values are not well observed and could further add to a hazard at that location due to its potential to go unseen. As visibility is a positive measure and the risk scores to date are negative measures, the visibility histogram needs normalisation and inverting (Eq. (6)). As a result the histogram, $\mathbf{H}_V$, then provides a measure of invisibility, within a range of zero to one.

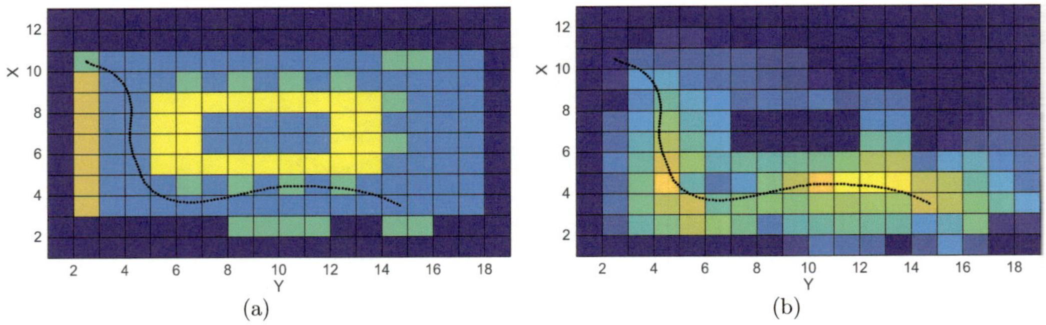

Figure 8. Visibility maps. (a) Individual path on which the visibility map is generated, (b) the visibility map produced from the agents field of view during the taken path.

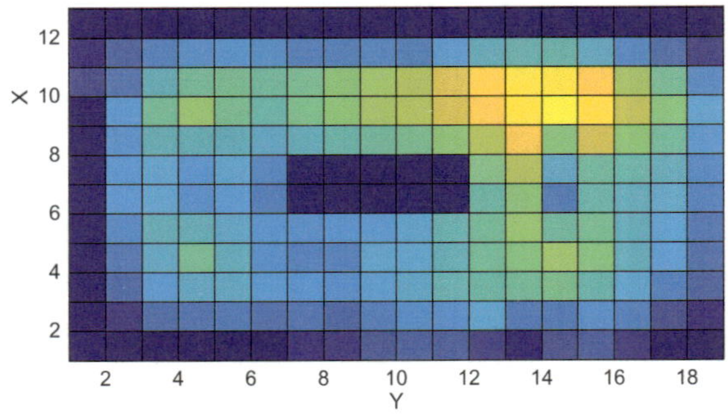

Figure 9. Resultant visibility map as a result of the simulated paths seen in **Figure 5b**.

$$\mathbf{H}_V = 1 - \frac{\mathbf{H}}{||\mathbf{H}||} \tag{6}$$

The final risk element E for the environmental risk maps is given as a combination of the risk associated with the presence of a person or robot and their ability to see hazards in a given environment. As such a weighted combination of the histograms produced by the interaction maps H_I and the visibility maps H_V is given as

$$E = w_V \mathbf{H}_V + w_I \mathbf{H}_I \tag{7}$$

where w represents the histograms contribution to the final risk element and the two contributions $w_V + w_I = 1$. The weighting is flexible based on the final implementation allowing the condition of those using the environment to be considered during simulation.

2.3. Simulation

To generate the position data used in the environmental risk maps, simulation algorithms are used to replicate the behaviours of those that interact with the scene. Simulating behaviour is often considered to have started with Reynolds work in 1987 with the emulation of animal behaviours [21]. Within Reynolds' work the concept of steering simulation was introduced, whereby each individual agent in a scene has its movement governed by a set of rules. In this case each agent follows three rules: steer towards the goal, steer away from the nearest obstacle, and steer away from the nearest person. Later Helbing et al. [22] introduced the Social Force Model (SFM) which uses potential fields defined by neighbouring agents to impart an acceleration to each agent. SFM's compute the trajectory of each agent by applying a series of forces to each agent that depend on the relative positions and velocities of nearby agents and the goal of the agent (Eq. (8)).

$$F_a = g_a - p_a + \sum_{i=1}^{n} f(a, b_i) + \sum_{j=1}^{m} f(o_j) + \sum_{k=1}^{o} f(a, b_k) \tag{8}$$

where g_a is the current destination of the agent a to its final goal, with p_a being the agent's current position. The forces for separation, $f(a, b_i)$, object avoidance $f(o_j)$ and predicative agent avoidance $f(a, b_k)$, is calculated for any relevant entity within a defined neighbourhood.

The simulation techniques described above relate to the local movement of agent from frame to frame. The overall path the agent follows through a scene is extrapolated at a higher level. Given a pre defined destination an agent performs a route plan using the A* algorithm to estimate the most direct course through the environment. A* is often used as it provides a near optimum path through the environment whilst being computationally efficient enough that it can be run on demand in real time if required [23]. Given that the intended applications for this work are likely indoor environments and may well be computed within the confines of a domestic robot, considerations as to the speed of runtime and algorithm applicability are important.

These simulation techniques govern local movement but not the overall strategy of the agents. With the integration of these factors, more rounded behaviour simulations began to be developed. The integration of risk related concepts into simulation algorithms has been a recent development, allowing for the prediction of crowd dynamics and behaviour in the presence of a stressor. This is especially important in areas such as architecture and engineering where through simulation more efficient evacuation plans and facilities can be built.

Kim et al. [24] focus on the modelling of the human perception of risk and how that subsequently effects behaviour. Using the general adaptation syndrome (GAS) model, proposed by Hans Selye (1956) [39], which acts as a model for the general response to any stressor (toxins, cold, injury, fatigue, fear, etc). The GAS model has three stages of response (**Figure 10**): alarm, the agent perceives a stressor and readies themselves for the 'fight' or 'flight' response. Resistance stage, in which agents work to resolve the stress at their full capacity and finally if the stressor is not resolved, they reach the exhaustion stage and resistance becomes futile.

Within this work, the goal is to approximate this model and apply the results of the various stages to the agents in a scene. This could range from mild stimulants, such as challenging situations (e.g. time constrained events) to threatening situations such as fire. These stressors are split into a number of categories from positional stressors, e.g. the effect on an agent from being in a certain risk area or the increasing effect of a stressor based on its proximity. Also the concept of interpersonal stressors is modelled, which allows the agent to perceive stress based on those around, for example crowding. **Figure 11** highlights the overview of this process,

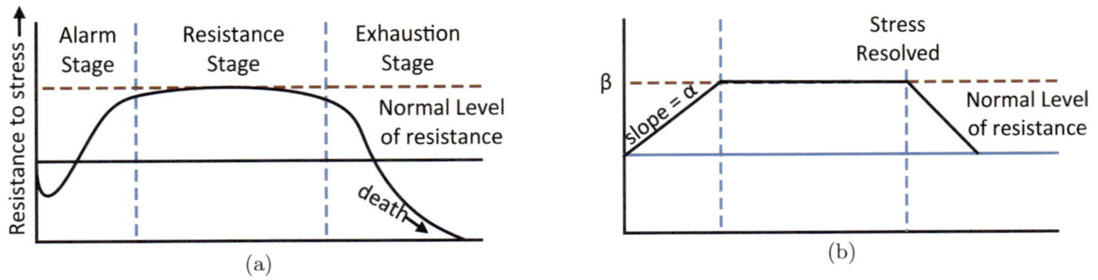

Figure 10. (a) The original General Adaptation Syndrome (GAS) model as described by Hans Selye (1956), (b) the approximation of the GAS model used to define agent behaviour.

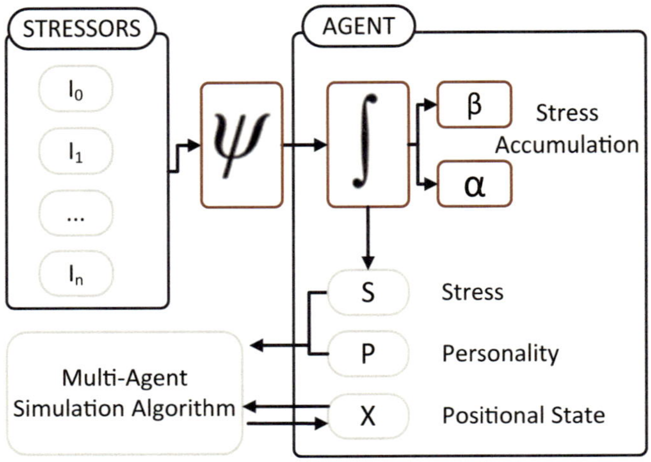

Figure 11. Overview of the simulation algorithm as proposed by Kim et al. [24].

where $I_1 \ldots n$ are the stressors. P represents an agents stable personality, which, given no stressors, would define the variables to drive the simulation algorithm. Ψ is the perceived stressor function. β and α control the agents stress response S, this defines how an agents responses will be effected in the presence of a stressors and is based on real world data. This directly impacts the way in which the agents act both at a personal and global level. Changing the way they react to others in the scene as well as driving decision making abilities regarding where they are going. Finally X represents the positional data of each agent and the resultant output of each frame of computation.

The addition of this stressor consideration allows for agents in a scenario to disregard common psychological behaviour effects that maintain coherent movement such as separation from those around. Instead the agent focuses on reaching their perceived goal as fast as possible. Through this addition the ability to model instances such as building evacuations under stress become more accurate. As a test case, the scenario of the Shibuya Crossing in Tokyo is utilised. In which agents are exposed to a mild time constrained event in which they must cross the road in the presence of many other agents and within a limited period. A quantitative assessment is made by looking at psychological study of pedestrians crossing roads [25], comparison is made of variables such as average crossing speeds against how much an agent has been delayed entering a street crossing from the start of the signal. This is then compared with the outputs of the simulation to provide a similarity measure.

Furthering this concept of modelling the perception of risk and how it effects behaviour, Klugl et al. [13] creates a model that aims to simulate an agents response to the evacuation of a train with an engine on fire in a tunnel. In this case rather than the perception of risk effecting their local movement, it is the decision making capabilities that are focused on. This is due to the need to understand what a agent would do in a given situation and allow the development of evacuation plans that compliment this. As such the cognitive model focuses on the perception of the risk, in this case fire, by the agents exposure to smoke and increasing temperature as well as other agents 'knowledge' of where the fire is located. Additionally factors such as visible signage and congestion, all effect the agent's strategic thinking for exiting the train tunnel. Exits are given as either the end of the tunnel or specific emergency exits. **Figure 12** demonstrates how the decision making components of the model effects the local movement, as well as an overview of how the decision making process is made.

A number of variables in the agent movement model are dedicated to the representation of risk to the agent. Panic variables are used to increase movement speed and reduce the agents concept of personal space. Although this is a less complex representation of the agents perception of risk, Klugl does provide a more in depth look at the conative decision making processes which allow for an agent to adjust their goals given the presence of a hazard and utilising available new information (**Figure 13**).

Using this focus on the higher level decision making process, utility and probability theory [26] can also be applied to the problem. Given that in a domestic environment the hazard situations that are encountered are likely to be less severe, the effect of panic on an agent is less of a priority. Instead focus should be on the reaction to the risk and whether that changes the way an agent interacts with environment. For example if a parent discovers a broken glass on the floor they will likely ensure that any children present will navigate through an environment differently or avoid it all together.

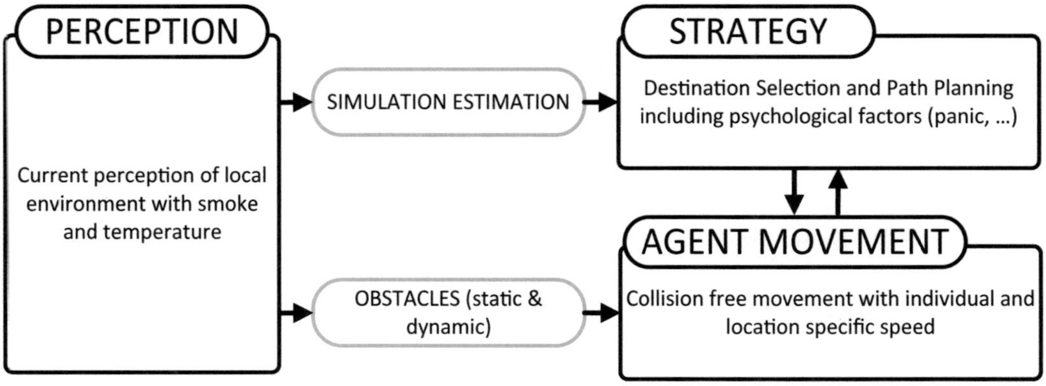

Figure 12. The high level description of the of the simulation algorithm proposed by Klugl et al. [13].

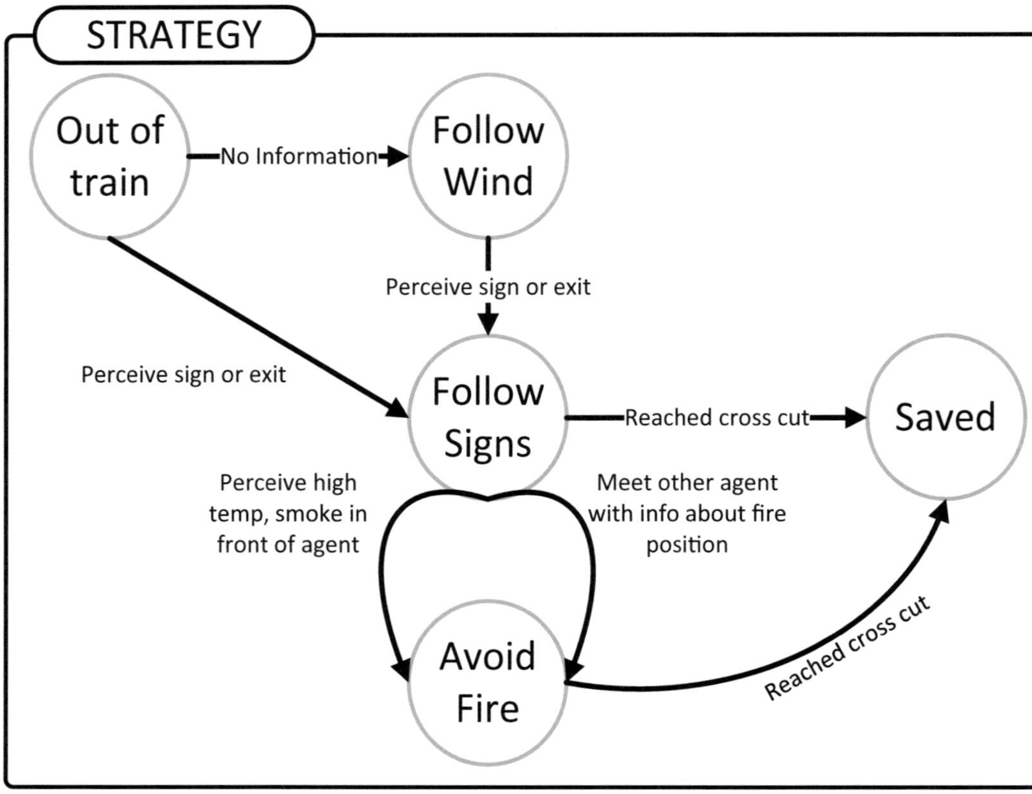

Figure 13. An in depth look at the strategy component of the method proposed by Klugl et al. [13].

Therefore by applying utility theory, predictions on how interactions change in the presence of risk can be made. In any given situation, the action an agent will take is defined by a set of probabilities. In the domestic situation the agent decides to either continue on their existing path or recalculate a new one to avoid a risk. As path finding and an agent's local movement are governed by other aspects of the simulation algorithm it is only this risk related choice that requires a decision making process.

Let **S** be a state at a given time in a simulation for an agent. **Figure 14** outlines this decision process for state **S** as a decision network. Within the diagram the decision node represents the problem to be resolved by a set of actions A. In this case which action to take whilst traversing the environment (continue or reroute). The chance nodes are indicative of the probabilistic outcomes associated with this decision represented as probability distribution functions. In this case the chance of injury and the chance of being able to find a path to the destination. The evidence nodes represent the knowledge the agent has, which directly affects the chance nodes. For example the risk of injury will change based on an agent seeing a hazard. Finally the utility node is the utility or value for the state **S** based on the chance nodes and an agent's preference.

To determine the best action for the agent to take, the principle of maximum expected utility (MEU) is utilised in 9. Here the possible actions A are assessed using the expected utility EU,

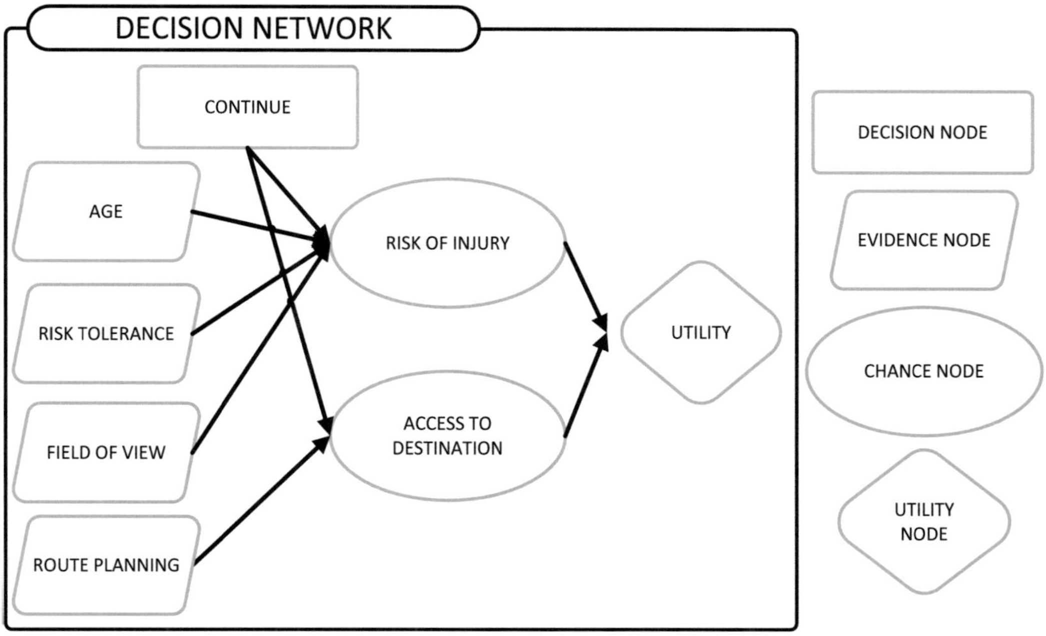

Figure 14. Decision network for risk interaction.

based on the known evidence **E**. Therefore the MEU represents the action A_i which is classified as most favourable by the agent.

$$MEU(A_i|\mathbf{E}) = \max \arg_A \sum_{i=1}^{a} EU(A_i|\mathbf{E}) \tag{9}$$

Evidence is a combination of agent variables and environmental feedback, for example for a human $\mathbf{E} = [p_a, p_r, p_p, p_f]$ where p_a is the age of the agent, p_r is their tolerance of risk (represented by Gaussian distributions, assuming prior knowledge of the means and standard deviations of the relevant information), p_p is the presence of an alternative route and p_f represents a seen hazard. In this simulation environment there are two actions that can be taken by an agent representing a single decision; continue on their preplanned path, or reroute to avoid the hazard.

The expected utility of an action is defined in 10. Expected utility represents a measure of both the likelihood of a particular state occurring, combined with the agent's preference for that outcome. Here a possible action A has a number of possible outcome states $\text{Result}_i(A)$. For each outcome a probability is assigned based on the evidence **E**. $\text{Do}(A)$ represents the supposition that the action A is executed in the current state.

$$EU(A|\mathbf{E}) = \mathbf{P}(\text{Result}_i(A)|\mathbf{E}, \text{Do}(A))U(\text{Result}_i(A)) \tag{10}$$

The given probability of each action A is then multiplied by a utility function U for the possible outcome states. In this example the utility function is simplistic as only a limited number of states exist based on the decision network presented in **Figure 14**. The utility associated with accessing the destination will nearly always take precedence; if the direct route to the destination is accessible then the agent will continue, only in cases where risk of injury is high will the agent decide to reroute. However if the route is blocked then regardless of the risk of injury the agent will have to reroute.

As an example, in a normal situation with the absence of risk, the probability associated with risk of injury is low and the probability for reaching the destination is high. Given that the agent wants to get to the destination, whilst avoiding injury, the $EU(A|\mathbf{E})$ for the A to carry on the current path is high. However if the evidence changes and a risk is detected through an agent's field of view, the probability of risk of injury increases. If there exists an additional route the agent can take to avoid the risk, the $EU(A|\mathbf{E})$ for the A to reroute will be higher and therefore the preferred option.

As a result of these simulation techniques a detailed representation of environment interaction can be created and used as the bases from which the environmental risk maps can be computed. However as with any simulation algorithm the evaluation of how realistic it is a complex problem, defining how realistic the simulation outputs are impacts the accuracy of the interactions maps, and therefore the accuracy of the produced risk evaluation. As such the topic of simulation evaluation is reviewed, providing a number of examples of how the validation of the simulation methods can be done.

2.4. Simulation evaluation

One of the most prominent issues with behaviour simulation research is the lack of a simple and suitable form of evaluation from source data to simulation output. This task is made more difficult as the developed simulation approaches cover a huge range of applications, where evaluation techniques for one are not always applicable to the others. Generally the evaluation techniques utilised can be split into qualitative [27] and quantitative measures [24, 28]. The former including assessments made by experts in the field or context of the intended application [13], as well as category based rating systems [29] designed to define the capabilities of an algorithm (such as emergent behaviours).

The ability to have a single value or set of values to define how good a simulation is, is a favourable goal and requires analysis of the outputs of the simulation algorithm. Kapadia et al. [30] look into the assessment of simulation algorithms using an exhaustive test case format. Evaluation of a algorithm is based on its ability to navigate 'steering problems', i.e. the traversal of a scene to a predefined goal, whilst avoiding both static and dynamic objects. Using procedurally generated steering problems, a specific simulation algorithm can be tested across a broad range of scenarios.

To facilitate this, two notions are presented: the first is the concept of scenario spaces, and secondly, metrics to quantify the coverage and the quality of a simulation algorithm in this space. Scenario space is defined as a set of parameters from which environments can be generated to test a simulation algorithm in. These values consist of upper and lower values for numbers of agents, obstacles and environment size amongst others. **Figure 15** demonstrates

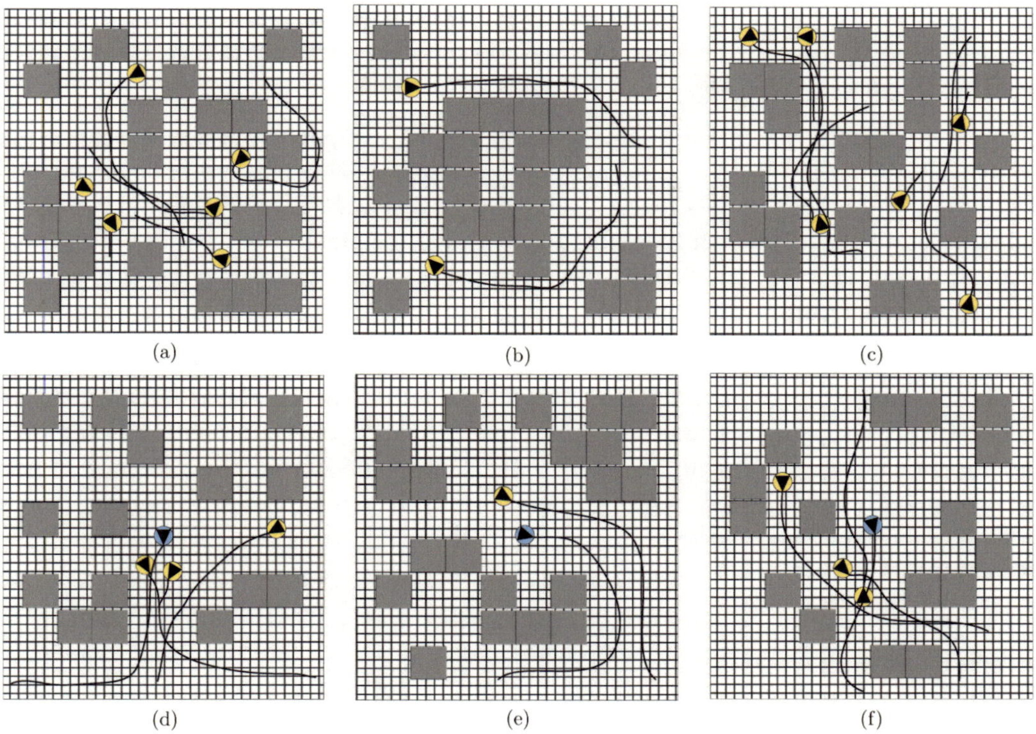

Figure 15. (a–c) Example steering problem scenarios, randomly generated from the scenario space. Black lines indicate the agents optimal path to their goal. (d–f) Steering problems generated in the representative scenario space. Sampling from this space ensures that all agents interact with the reference agent (in blue) which is always located at the centre of the environment.

some example steering problems taken from a scenario space. A successful generated space must adhere to a set of rules, for example an agent must be able to get from its origin to its goal and, on initialisation, there should be no collisions between agents or objects. A successfully completed scenario is defined as a boolean value, completed or not. To meet this a simulation algorithm must navigate an agent through the environment within a fixed time and without colliding with other agents or static objects. This process is repeated for a representative number of possible combinations derived from the scenario space.

Evaluation of the successful spaces is done based on three metrics: scenario completion, length of time and distance travelled. These are compared to optimal values calculated when generating that specific test space. Using these metrics, concepts of coverage, quality and failure are computed for a given simulation algorithm for a given scenario space. By substituting different algorithms and using the same scenario spaces, comparisons between algorithm performance can be established.

This exhaustive form of analysis is useful in testing the limits of a given simulation algorithm and highlighting types of scenario where a particular method might be improved. In addition the automated nature of the test generation provides a useful platform from which easy benchmarking can be performed. However, the exhaustive randomly generated scenario spaces that are created lack similarity to real world scenarios nor is any real comparison made with source data.

As such the next logical approach is the development of a comparative system which can use source data as a comparison or benchmark from which to evaluate the output of simulation algorithms. This could provide overall similarity ratings but also more localised analysis, whereby areas of interest or high deafferentation could be highlighted specifically. Alternatively the simulation outputs themselves can be analysed for outliers, highlighting possible anomalous data relative to the rest of the outputs.

One such example is the work of Charalambous et al. [31] in which a data driven approach is applied to the problem by developing an analytics framework for anomaly detection. Unusual behaviour is principally presented using two processes. Firstly outlier detection which takes a set of data and analyses it with no other reference data, allowing for the definition of odd behaviour within just that dataset. Secondly novelty detection, in which sample data is compared to reference material to find and describe trends or actions that differ from the reference data.

Using segment analysis of agent trajectories, a set of metrics are used to characterise each portion of the agent track, these include properties such as average speed, maximum curvature or minimum distance to nearest neighbour. Using this set of metrics, a nominal data representation is created for that dataset, this can be used as training for similar unseen data. Which metrics are used to create this representation of nominal data is user defined and will vary based on the types of data and the scenario. Using techniques such as Once Class SVM, Localised p-value estimation (*k*-LPE) and *k*-NN Based Approaches, outliers in the test data can be highlighted.

Off the back of this analysis a set of user-in-the-loop analysis tools are created, which provide an interface by which to interpret the data. Due to the low level analysis of the simulation this can localise specific agents that are acting erroneously or where general areas of inconsistency appear in the test scenes.

Another alternative to data driven evaluation approaches is through the use of visual comparative frameworks. This builds on the notion that a simulation's accuracy is not entirely based on its ability to reproduce specific agents movement exactly, rather the concept of similar looking simulations are considered better. This provides a more general approach to evaluation and replicated better the natural formation of crowds. For example people moving through a train station day to day will follow similar paths, but will experience very slightly different crowd dynamics, likely causing them to vary their path slightly from the previous day. This would represent a change in behaviour but not necessarily an anomalous one. Regimented data driven techniques do not allow for this variation and ignore whether the crowd movement looks similar.

The following framework makes use of compositing techniques and video analysis tools to compare source footage to simulated. Evaluation can be done on a frame by frame basis or on a sequence as a whole, providing flexibility in how the simulation is evaluated. Additionally the methodology requires no track or path information from the source material, allowing any video footage captured from a static viewpoint to be used.

Comparison is made using the original source footage and a video created using composition techniques, in which simulated agents are superimposed into the background of the source video data (**Figure 16**). A set of metrics designed to evaluate the visual similarity of the two videos is used to provide a quantifiable similarity metric. These are designed to emulate the way the human visual system (HVS) perceives motion, both in direction and volume. Fundamentally the framework is made up of two components; simulation visualisation and video similarity.

Figure 17 provides a brief overview of the process. As the framework compares video data to derive a similarity value, firstly a simulated video must be constructed. Using the source video sequence, the background is extracted. A two dimensional plane representing a top down view of the given scene is used as the simulation space. Simulations are run to produce paths for virtual agents to follow based on the extracted plane. The visualisation component is used to composite the extracted 2D background image and 3D rendered agents as they follow the simulated paths. Frames are output from the visualisation into a final simulated video sequence (**Figure 16**). Once both a simulated and source video are available, the similarity can be evaluated. Optical flow and tracklet analysis are run and features extracted from the subsequent data. Finally a distance measure is used to analyse the difference in features to give the final similarity metric.

The visualisation stage of the framework performs the composition of a scene. Firstly the background of the source video is extracted using gaussian mixture models [32]. Next a perspective plane is generated in 3D based on a 2D mapping of the scene, this could be obtained during the environmental maps evaluation. The key to a visually similar composition is the positioning of a virtual camera at the same location as in the original scene, relative to the perspective plane. By using layers the camera can have the source image as a background and

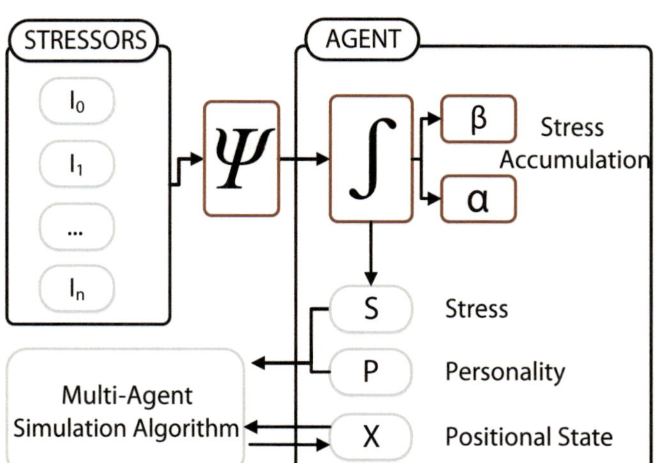

Figure 16. Frames of source CCTV footage and generated video using the composition techniques.

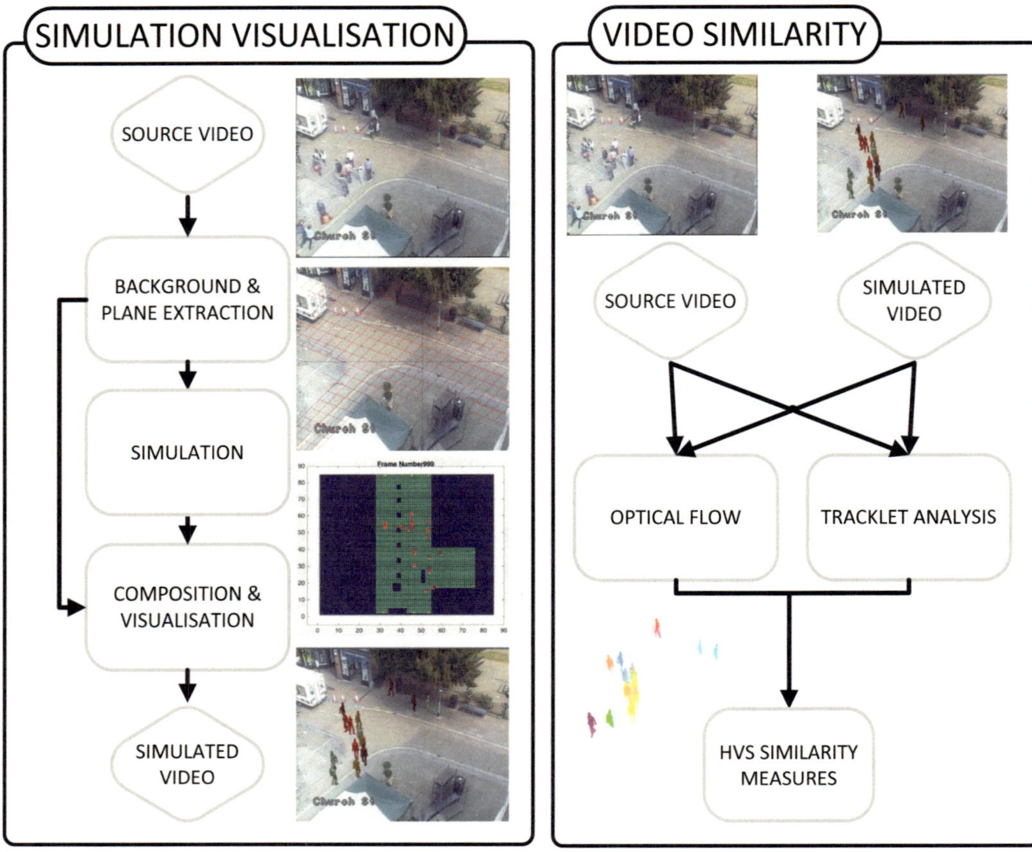

Figure 17. Overview of the human and group behaviour simulation evaluation framework.

visualised 3D agents, controlled by the simulation, superimposed. When correctly aligned the illusion of the 3D agents walking through the scene is created. This alignment can be performed manually using the position and orientation of the camera or automatically using camera calibration techniques [33, 34]. The 3D agents in the scene must also be scaled correctly to ensure their sizing is appropriate to the scene (**Figure 18**), this can be done manually or calculated automatically using the methods provided in Ref. [35].

To produce the simulated video, the 3D agents navigate through the scene with the virtual camera capturing composite images. For each frame of the visualisation the agent location and rotation is updated based on the simulation algorithm output. Once visualisation is completed the individual frames are compiled into a video sequence.

Once the composite video sequence is complete, the source and the simulated video sequences are used to extract human visual system's (HVS) features in order to measure their level of similarity. These features are based on the optical flow [36], Histogram of Oriented Optical Flow (HOOF) [37] and tracklets [38] of the moving objects in both sequences and are designed to emulate the way in which humans perceive motion. The feature representations used highlight different properties of video sequences allowing for a specific analysis of a simulations output,

Figure 18. Example composition scene with test agents and perspective floor plan.

for example tracklet analysis allows for the pathfinding component to be evaluated, similarly the histogram of optical flow provides a good analysis of crowd volume and speed.

In order to evaluate the similarity level of the simulated and source videos using the HVS features, a metric is required that will allow an objective comparison of the extracted features. The Bhattacharyya distance measure is utilised, providing an individual score of similarity between each video sequence across the features used. This can be expanded to a frame by frame, or even a block by block analysis, allowing spatio-temporal adaptation of the metrics. With this utility the final distance metric can highlight area of the video sequence which seem dissimilar from the source. Additionally through the comparison of individual features the type of analogy can be deduced.

3. Conclusion

The task of automated risk assessment in a domestic environment is an emerging area of research. The issues surrounding the identification, classification and, finally, quantification of risk are substantial. There are the practical issues of identifying a risk or hazard, as well as the contextual problem of defining what is considered as hazardous and what is not, both of which are non-trivial tasks. One such issue is the effect that human interaction has on detected risks in an environment and conversely the effect those risks have on those within the environment. More importantly methods are required to predict these effects and therefore provide further insight into those detected risks, allowing for a more complete picture of risk in an environment.

An introduction to risk as a function of interaction has been given in the form of environmental risk maps, which quantify presence and visibility to provide an risk element for use in the risk estimation framework. Environmental risk maps rely on the accuracy of the simulation algorithm used to produce realistic behaviour in an environment. This forms the basis from which the interaction and visibility components of the environmental risk maps are calculated, ensuring that outputted risk scores are logical and relevant.

As has been demonstrated the definition of simulation algorithms and models for predicting behaviour is a challenge, primarily due to the need for careful consideration when defining which aspects of human or robotic behaviour to model. This issue is exacerbated when modelling behaviour in the presence of risk. Crucially this simulation data is utilised for the definition of environmental risk maps, and therefore the accuracy of the simulation must be verified to ensure that the produced behaviour and by extension the produced risk scores are realistic. There is ongoing development in this field for the parametric definition of behaviour, i.e. which elements of behaviour effect the way we navigate and move? Extension of these parameters would allow for a more tailored approach to the simulation of agents in a scene, allowing more specific emulation of robotic devices or humans with disabilities or limitations, however this requires a detailed analysis of which factors need to be considered and how applicable they are.

A number of simulation techniques have been reviewed looking at the various aspects of behaviour modelling from movement mechanics up to the strategic decision making functionally. Finally as validation of simulation accuracy is vital to the relevance of the environmental risk maps a number of simulation algorithm evaluation techniques have been reviewed. A broad introduction to the topic has been provided in the form of exhaustive procedural analysis techniques through to outlier detection and data driven comparisons, finally a novel composition based analysis framework emulating a human's ability to identify similarities in movement is also reviewed.

Domestic robotics and smart homes are a growing industry and will become an integral part of life in the near future. Currently available commercial products aim to perform menial tasks, simplifying processes that humans perform every day. For example taking notes, the initialisation of domestic appliances, information retrieval and simple household chores. However with the ever more interconnected nature of the domestic environment and the availability of increasing computational power, these devices will take on new roles. The ability to provide basic decision making capabilities as well as more detailed interaction and analytical abilities will enable the application of more complex behaviours in these devices.

In a domestic setting this will likely lead to the performing of more complex tasks such as assisting with a wider range of household chores, heavy lifting and entertainment. However for scenarios involving the presence or potentially the monitoring of those that use the environment (e.g. children or at risk adults), elements of risk detection will be required to ensure user as well as device safety. These concepts will be required in the emulation of further higher level behaviours and will produce an initial step in the development of systems that can make a real difference in easing some of the impending social issues to do with health care and ageing populations. However with these developments also comes a vital need for the establishment of independent advisory and certification boards. These organisations would provide governance on the usability and appropriate audience of the various devices and in doing so provide minimum safety requirements for use with humans as well as practical guidelines for device preservation.

Author details

Rob Dupre* and Vasileios Argyriou

*Address all correspondence to: r.dupre@kingston.ac.uk

Kingston University, UK

References

[1] White MG. Kitchen Hazards; 2014. LoveToKnow. Retrieved April 10, 2017, from http://safety.lovetoknow.com/Kitchen_Hazards

[2] Office for National Statistics. Disability in England and Wales: 2011 and Comparison with 2001; 2013. Office for National Statistics. Retrieved April 5, 2017 from http://www.ons.gov.uk/peoplepopulationandcommunity/healthandsocialcare/disability/articles/disabilityinenglandandwales/2013-01-30

[3] Statistics Of N. 2011 Census: Population and Household Estimates for the United Kingdom; 2012. Office for National Statistics. Retrieved April 5, 2017 from http://webarchive.nationalarchives.gov.uk/20160105160709/http://www.ons.gov.uk/ons/guide-method/census/2011/census-data/2011-census-data-catalogue/population-and-household-estimates/index.html

[4] The Royal Society for the Prevention of Accidents. HASS and LASS Data; 2002. Home & Leisure Accident Surveillance System. Retrieved March 20, 2017 from http://www.hassandlass.org.uk/reports/2002data.pdf%5Cnhttp://www.hassandlass.org.uk/query/MainSelector.aspx

[5] Melorose J, Perroy R, Careas S. World Population Prospects; 2015. United Nations: Department of Economic and Social Affairs. Retrieved April 5, 2017 from https://esa.un.org/unpd/wpp/publications/files/key_findings_wpp_2015.pdf

[6] International Federation of Robotics. World 2014 Robotics Survey—executive summary. World Robotic Report—Executive Summary. 2014: pp. 10–21

[7] Kobayashi T, Aoyama T, Sekiyama K, Fukuda T. Selection algorithm for locomotion based on the evaluation of falling risk. IEEE Transactions on Robotics. 2015;**31**(3):750–765

[8] Dupre R, Argyriou V. 3D voxel HOG and risk estimation. In: International Conference on Digital Signal Processing. DSP. Vol. 2015-Sep;2015. pp. 482–486

[9] Adali T, Schreier PJ, Scharf LL. Complex-valued signal processing: The proper way to deal with impropriety. IEEE Transactions on Signal Processing. 2011;**59**(11):5101–5125

[10] Zheng B, Zhao Y, Yu J, Ikeuchi K, Zhu SC. Scene understanding by reasoning stability and safety. International Journal of Computer Vision. 2015;**112**(2):221–238

[11] Dupre R, Argyriou V, Greenhill D. A 3D Scene Analysis Framework and Descriptors for Risk Evaluation. In: International Conference on 3D Vision (3DV). IEEE; 2015. pp. 100–108

[12] Helbing D, Farkas I, Vicsek T. Simulating dynamical features of escape panic. Nature. 2000;**407**(6803):487–490

[13] Klugl F, Klubertanz G, Rindsfuser G. Agent-based pedestrian simulation of train evacuation integrating environmental data. Lecture Notes in Computer Science. 2009;**5803**:631–638

[14] Dupre R, Argyriou V, Tzimiropoulos G, Greenhill D. Risk analysis for smart homes and domestic robots using robust shape and physics descriptors, and complex boosting techniques. Information Sciences. 2016;**372**:359–379

[15] Younes G, Asmar D, Shammas E. A survey on non-filter-based monocular visual SLAM systems. 2016; arXiv:160700470

[16] Fuentes-Pacheco J, Ruiz-Ascencio J, Rendon-Mancha JM. Visual simultaneous localization and mapping: A survey. Artificial Intelligence Review. 2012;**43**(1):55–81

[17] Newcombe Ra, Davison AJ, Izadi S, Kohli P, Hilliges O, Shotton J. et al. KinectFusion: Real-time dense surface mapping and tracking. In: 2011 10th IEEE International Symposium on Mixed and Augmented Reality; Oct 2011; pp. 127–136

[18] Izadi S, Davison A, Fitzgibbon A, Kim D, Hilliges O, Molyneaux D. et al. Kinect Fusion: Real-time 3D reconstruction and interaction using a moving depth camera. In: Proceedings of the 24th Annual ACM Symposium on User Interface Software and Technology; 2011. pp. 559–568

[19] Fernandez-Carames C, Moreno V, Curto B, Rodriguez-Aragon JF, Serrano FJ. A real-time door detection system for domestic robotic navigation. Journal of Intelligent and Robotic Systems: Theory and Applications. 2013;**76**(1):119–136

[20] Dragoi V. Chapter 14: Visual processing: Eye and retina. In: Concha J, editor. Neuroscience Online, the Open-Access Neuroscience Electronic Textbook. University of Texas Medical School at Houston. 1997: pp. 1–16

[21] Reynolds CW. Flocks, herds and schools: A distributed behavioral model. In: ACM SIGGRAPH Computer Graphics. Vol. 21. ACM. 1987; pp. 25–34

[22] Helbing D, Molnar P. Self-organization phenomena in pedestrian crowds. Condensed Matter. 1998:569–577

[23] Foudil C, Noureddine D, Sanza C, Duthen Y. Path finding and collision avoidance in crowd simulation. Journal of Computing and Information Technology. 2009;**17**(3): 217–228

[24] Kim S, Guy SJ, Manocha D, Lin MC. Interactive simulation of dynamic crowd behaviors using general adaptation syndrome theory. ACM SIGGRAPH Symposium on Interactive 3D Graphics and Games—I3D. 2012;**1**(212):55

[25] Crompton D. Pedestrian delay, annoyance and risk: Pre-liminary results from a 2 years study. In: Proceedings of PTRC Summer Annual Meeting; 1979. pp. 275–299

[26] Russell SJ, Norvig P. Artificial Intelligence: A Modern Approach; Neurocomputing 1995. p. 9

[27] Portz A, Seyfried A. Analyzing stop-and-go waves by experiment and modeling. Pedestrian and Evacuation Dynamics. Springer. 2010;577–586

[28] Asano M, Iryo T, Kuwahara M. A pedestrian model considering anticipatory behaviour for capacity evaluation. Transportation and Traffic Theory. 2009;**18**:28

[29] Duives DC, Daamen W, Hoogendoorn SP. State-of-the-art crowd motion simulation models. Transportation Research Part C: Emerging Technologies. 2013;**37**:193–209

[30] Kapadia M, Wang M, Singh S, Reinman G, Faloutsos P. Scenario space: Characterizing coverage, quality, and failure of steering algorithms. Proceedings of the 2011 ACM SIGGRAPH/Eurographics Symposium on Computer Animation—SCA '11. Vol. 1; 2011. p. 53

[31] Charalambous Pa, Karamouzas Ib, Guy SJb, Chrysanthou Ya. A data-driven framework for visual crowd analysis. Computer Graphics Forum. 2014;**33**(7):41–50

[32] Zivkovic Z. Improved adaptive Gaussian mixture model for background subtraction. In: Proceedings of the 17th International Conference on Pattern Recognition. 2004; ICPR 2004. vol. 2; 2004. pp. 28–31

[33] Guo S, Qu Z, Wang L. Camera pose estimation using frequency analysis. In: 2014 International Conference on Information Science and Applications (ICISA). 2014: pp. 3–6

[34] Zeisl B, Sattler T, Pollefeys M. Camera pose voting for large-scale image-based localization. Proceedings of the IEEE International Conference on Computer Vision. 2015: pp. 2704–2712

[35] Do Y. On the neural computation of the scale factor in perspective transformation camera model. In: IEEE International Conference on Control and Automation (ICCA). 2013. pp. 712–714

[36] Sun D, Roth S, Black MJ. Secrets of optical flow estimation and their principles. Proceedings of the IEEE Computer Society Conference on Computer Vision and Pattern Recognition; 2010. pp. 2432–2439

[37] Chaudhry R, Ravichandran A, Hager G, Vidal R. Histograms of oriented optical flow and Binet-Cauchy kernels on nonlinear dynamical systems for the recognition of human actions. In: 2009 IEEE Computer Society Conference on Computer Vision and Pattern Recognition Workshops (CVPR Workshops 2009); 2009. pp. 1932–1939

[38] Khoong WL, Kow WY, Tan HT, Yoong HP, Teo K, Tze K. Kalman filtering based object tracking in surveillance video system. In: Proceedings of the 3rd CUTSE International Conference; 2011

[39] Selye H. The stress of life. No. v. 5 in McGraw.

Chapter 3

Trajectory Tracking Error Using Fractional Order PID Control Law for Two-Link Robot Manipulator via Fractional Adaptive Neural Networks

Joel Perez P., Jose Paz Perez P. and
Martha S. Lopez de la Fuente

Additional information is available at the end of the chapter

http://dx.doi.org/10.5772/intechopen.70020

Abstract

The problem of trajectory tracking of unknown nonlinear systems of fractional order is solved using fractional order dynamical neural networks. For this purpose, we obtained control laws and laws of adaptive weights online, obtained using the Lyapunov stability analysis methodology of fractional order. Numerical simulations illustrate the obtained theoretical results.

Keywords: fractional order PID control, fractional adaptive neural networks, fractional Lyapunov functions, fractional nonlinear systems, trajectory tracking

1. Introduction

The fractional calculus is a branch of mathematics that attracted attention since G.W. Leibnitz proposed it in the seventeenth century. However, the researchers were not attracted to this area because of the lack of applications and analytical results of the fractional calculus.

On the contrary, the fractional calculus currently attracts the attention of a large number of scientists for their applications in different fields of science, engineering, chemistry, and so on.

This chapter presents the design of a fractional order nonlinear identifier modeled by a dynamic neural network of fractional order.

Although PID controllers are introduced long time ago, they are widely used in industry because of their advantages such as low price, design simplicity, and suitable performance. While three parameters of design including proportional (Kp), integral (Ki), and derivative

(Kd) are available in PID controllers, two more parameters exist in FOPID controllers for adjustment. These parameters are integral fractional order and derivative fractional order. In comparison with PID controllers, FOPID controllers have more flexible design that results in more precise adjustment of closed-loop system. FOPID controllers are defined by FO differential equations. It is possible to tune frequency response of the control system by expanding integral and derivative terms of the PID controller to fractional order case. This characteristic feature results in a more robust design of control system, but it is not easily possible. According to nonlinearity, uncertainty, and confusion behaviors of robot arms, they are highly recommended for experimenting designs of control systems. Despite nonlinear behavior of robot arm, it is demonstrable that a linear proportional derivative controller can stabilize the system using Lyapunov. But, classic PD controller itself cannot control robot to reach suitable condition. Several papers and wide researches in optimizing performance of the robot manipulator show the importance of this issue.

There are several ways of defining the derivative and fractional integral, for example, the derivative of Grunwald-Letnikov given by Eq. (1)

$$aD_t^\alpha f(t) = \lim_{h\to 0}\frac{1}{h^\alpha}\sum_{j=0}^{\left[\frac{(t-\alpha)}{h}\right]}(-1)^j\binom{\alpha}{j}f(t-jh) \tag{1}$$

where [.] is a flooring operator, while the RL definition is given by:

$$aD_t^\alpha f(t) = \frac{1}{\Gamma(n-\alpha)}\frac{d^n}{dt^n}\int_a^t \frac{f(\tau)}{(t-\tau)^{\alpha-n+1}}d\tau \tag{2}$$

For $(n-1<\alpha<n)$, $\Gamma(x)$ is the well-known Euler's Gamma function.

Similarly, the notation used in ordinary differential equations, we will use the following notation, Eq. (3), when we are referring to the fractional order differential equations where $\alpha_k \in \mathbb{R}^+$.

which is:

$$g(t,\ x, aD_t^{\alpha_1}\ x,\ aD_t^{\alpha_2}\ x, \dots\ ) = 0 \tag{3}$$

The Caputo's definition can be written as

$$aD_t^\alpha f(t) = \frac{1}{\Gamma(\alpha-n)}\int_a^t \frac{f^{(n)}(\tau)}{(t-\tau)^{\alpha-n+1}}d\tau \tag{4}$$

For

$$(n-1<\alpha<n).$$

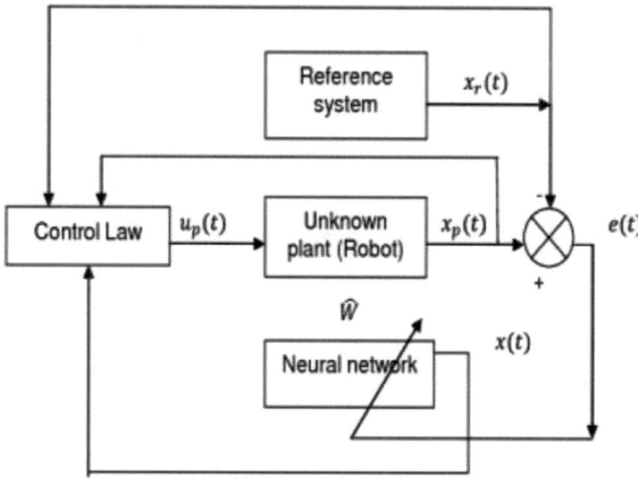

Figure 1. Recurrent neural network scheme.

Trajectory tracking, synchronization, and control of linear and nonlinear systems are a very important problem in science and control engineering. In this chapter, we will extend these concepts to force the nonlinear system (plant) to follow any linear and nonlinear reference signals generated by fractional order differential equations.

The proposed adaptive control scheme is composed of a recurrent neural identifier and a controller (**Figure 1**).

We use the above scheme to model the unknown nonlinear system by means of a dynamic recurrent neural network of adaptable weights; the above is modeled by differential equations of fractional order. Also, the scheme allows us to determine the control actions, the error of approach of trajectories, as well as the laws of adaptation of adaptive weights and the inter-connection of such systems.

2. Modelling of the plant

The nonlinear system (Eq. (5)) is forced to follow a reference signal:

$$aD_t^{\alpha}x_p = F_p(x_p, u) \triangleq f_p(x_p) + g_p(x_p)u_p \tag{5}$$

$$x_p,\ f_p. \in \mathbb{R}^n,\ u \in \mathbb{R}^m,\ g_p \in \mathbb{R}^{nxn}.$$

The differential equation will be modeled by:

$$aD_t^{\alpha}x_p = \mathrm{A}(x) + \mathrm{W}^*\Gamma_z(x) + \Omega_u.$$

The tracking error between these two systems:

$$w_{per} = x - x_p \tag{6}$$

We use the next hypotheses.

$$aD_t^{\alpha} w_{per} = -k w_{per} \tag{7}$$

In this research, we will use $k = 1$, so that, Eq. (6), $aD_t^{\alpha} w_{per} = aD_t^{\alpha} x - aD_t^{\alpha} x_p$,; so

$$aD_t^{\alpha} x_p = aD_t^{\alpha} x + w_{per}$$

The nonlinear system is [1]:

$$aD_t^{\alpha} x_p = aD_t^{\alpha} x + w_{per} = \mathrm{A}(x) + \mathrm{W}^* \Gamma_{\mathrm{z}}(x) + w_{per} + \Omega_{\mathrm{u}} \tag{8}$$

Where the W^* is the matrix weights.

3. Tracking error problem

In this part, we will analyze the trajectory tracking problem generated by

$$aD_t^{\alpha} x_r = f_r(x_r, u_r), \quad w_r, \ x_r \in \mathbb{R}^n \tag{9}$$

Are the state space vector, input vector, and f_r, is a nonlinear vectorial function.

To achieve our goal of trajectory tracking, we propose

$$e = x_p - x_r \tag{10}$$

The time derivative of the error is:

$$aD_t^{\alpha} e = aD_t^{\alpha} x_p - aD_t^{\alpha} x_r = \mathrm{A}(x) + \mathrm{W}^* \Gamma_{\mathrm{z}}(x) + w_{per} + \Omega_{\mathrm{u}} - f_r(x_r, u_r) \tag{11}$$

The Eq. (11) can be rewritten as follows, adding and subtracting the next terms $\widehat{W}\Gamma_{\mathrm{z}}(x_r)$, $\alpha_r\left(\mathrm{t}, \widehat{W}\right)$, Ae and $w_{per} = x - x_p$,; then,

$$\begin{aligned} aD_t^{\alpha} e &= \mathrm{A}(x) + \mathrm{W}^* \Gamma_{\mathrm{z}}(x) + x - x_p + \Omega_{\mathrm{u}} - f_r(x_r, u_r) + \widehat{W}\Gamma_{\mathrm{z}}(x_r) \\ &\quad - \widehat{W}\Gamma_{\mathrm{z}}(x_r) + \Omega\alpha_r\left(\mathrm{t}, \widehat{W}\right) - \Omega\alpha_r\left(\mathrm{t}, \widehat{W}\right) + Ae - Ae \\ aD_t^{\alpha} e &= \mathrm{A}e + \mathrm{W}^* \Gamma_{\mathrm{z}}(x) + \Omega_{\mathrm{u}} - f_r(x_r, u_r) + \widehat{W}\Gamma_{\mathrm{z}}(x_r) + \Omega\alpha_r\left(\mathrm{t}, \widehat{W}\right) \\ &\quad - \widehat{W}\Gamma_{\mathrm{z}}(x_r) - \Omega\alpha_r\left(\mathrm{t}, \widehat{W}\right) - Ax_r - x_r + x + A(x) \end{aligned} \tag{12}$$

The unknown plant will follow the fractional order reference signal, if:

$Ax_r + \widehat{W}\Gamma_{\mathrm{z}}(x_r) + x_r - x_p + \Omega\alpha_r\left(\mathrm{t}, \widehat{W}\right) = f_r(x_r, u_r)$, where

$$\Omega\alpha_r\left(t,\widehat{W}\right) = f_r(x_r,u_r) - Ax_r - \widehat{W}\Gamma_z(x_r) - x_r + x_p \tag{13}$$

$$aD_t^\alpha e = Ae + W^*\Gamma_z(x) - \widehat{W}\Gamma_z(x_r) - Ae + (A+I)(x-x_r) + \Omega\left(u - \alpha_r\left(t,\widehat{W}\right)\right) \tag{14}$$

Now, $\widehat{W}$ is part of the approach, given by W^*. The Eq. (14) can be expressed as Eq. (15), adding and subtracting the term $\widehat{W}\Gamma_z(x)$ and if $\Gamma_z(x) = \Gamma(z(x) - z(x_r))$

$$aD_t^\alpha e = Ae + \left(W^* - \widehat{W}\right)\Gamma_z(x) + \widehat{W}\Gamma(z(x) - z(x_r)) + (A+I)(x-x_r) - Ae + \Omega\left(u - \alpha_r\left(t,\widehat{W}\right)\right) \tag{15}$$

If

$$\widetilde{W} = W^* - \widehat{W} \text{ and } \widetilde{u} = u - \alpha_r\left(t,\widehat{W}\right) \tag{16}$$

And by replacing Eq. (16) in Eq. (15), we have:

$$aD_t^\alpha e = Ae + \widetilde{W}\Gamma_z(x) + \widehat{W}\Gamma(z(x) - z(x_r)) + (A+I)(x-x_r) - Ae + \Omega\widetilde{u}$$

$$\begin{aligned} aD_t^\alpha e = Ae + \widetilde{W}\Gamma_z(x) + \widehat{W}\Gamma\left(z(x) - z\left(x_p\right) + z\left(x_p\right) - z(x_r)\right) + \\ (A+I)\left(x - x_p + x_p - x_r\right) - Ae + \Omega\widetilde{u} \end{aligned} \tag{17}$$

And:

$$\widetilde{u} = u_1 + u_2 \tag{18}$$

So, the result for Ωu_1 is

$$\Omega u_1 = -\widehat{W}\Gamma\left(z(x) - z\left(x_p\right)\right) - (A+I)\left(x - x_p\right), \tag{19}$$

and Eq. (17) is simplified:

$$aD_t^\alpha e = Ae + \widetilde{W}\Gamma_z(x) + \widehat{W}\Gamma\left(z\left(x_p\right) - z(x_r)\right) + (A+I)\left(x_p - x_r\right) - Ae + \Omega\widetilde{u}$$

Taking into account that $e = x_p - x_r$, the equation for aD_t^αe is

$$\begin{aligned} aD_t^\alpha e &= (A+I)e + \tilde{W}\Gamma_z(x) + \widehat{W}\Gamma(z(e+x_r) - z(x_r)) + \Omega u_2 \\ &= (A+I)e + \widetilde{W}\sigma(x) + \widehat{W}(\sigma(e+x_r) - \sigma(x_r)) + \Omega u_2 \end{aligned}$$

If $\phi(e) = (\sigma(e+x_r) - \sigma(x_r))$, then

$$aD_t^\alpha e = (A+I)e + \widetilde{W}\sigma(x) + \widehat{W}\phi(e) + \Omega u_2 \tag{20}$$

Now, the problem is to find the control law Ωu_2, in which it stabilizes to the system Eq. (20). We will obtain the control law using the fractional order Lyapunov methodology.

4. Asymptotic stability of the approximation error

From Eq. (20), we consider the stability of the tracking error, for which we first observe that $\left(e, \widehat{W}\right) = 0$, is an equilibrium state of dynamical system from Eq. (20), and we consider a particular case when $A = -\lambda I, \lambda > 0$

For such stability analysis of the trajectory tracking (Eq. (20)), we propose the following FOPID control law [2]:

$$\Omega u_2 = K_p e + K_i a D_t^{-\alpha} e + K_v a D_t^{\alpha} e - \gamma\left(\frac{1}{2} + \frac{1}{2}\|\widehat{W}\|^2 L_{\phi}^2\right) e \tag{21}$$

Our objective is to find K_p, K_i, K_v, $\widehat{W}$, $L_{\phi_z}^2$, and this guarantees that the tracking error given by Eq. (20) is asymptotically stable, for which we will later propose a Lyapunov function, with $\gamma > 0$; this control law (Eq. (21)) is similar to [3].

A FOPID controller, also known as a $PI^{\lambda} D^{\alpha}$ controller, takes on the form [4]:

$$u(t) = K_p e(t) + K_i a D_t^{-\lambda} e(t) + K_d a D_t^{\alpha} e(t)$$

where λ and α are the fractional orders of the controller and $e(t)$ is the system error, where $\lambda = \alpha$. Note that the system error $e(t)$ replaces the general function $f(t)$.

We will show that the feedback system is asymptotically stable. Replacing Eq. (21) in Eq. (20), we have

$$aD_t^{\alpha} e = (A + I)e + \widetilde{W}\sigma(x) + \widehat{W}\phi(e) + K_p e + K_i a D_t^{-\alpha} e + K_v a D_t^{\alpha} e - \gamma\left(\frac{1}{2} + \frac{1}{2}\|\widehat{W}\|^2 L_{\phi}^2\right) e, \text{ then}$$

$$(1 - K_v) a D_t^{\alpha} e = (A + I)e + \widetilde{W}\sigma(x) + \widehat{W}\phi(e) + K_p e + K_i a D_t^{-\alpha} e - \gamma\left(\frac{1}{2} + \frac{1}{2}\|\widehat{W}\|^2 L_{\phi}^2\right) e. \text{ If}$$

$a = (1 - K_v)$, then

$$aD_t^{\alpha} e = \frac{1}{a}(A + I)e + \frac{1}{a}\widetilde{W}\sigma(x) + \frac{1}{a}\widehat{W}\phi(e) + \frac{1}{a}K_p e + \frac{1}{a}K_i a D_t^{-\alpha} e - \frac{\gamma}{a}\left(\frac{1}{2} + \frac{1}{2}\widehat{W}^2 L_{\phi}^2\right) e \tag{22}$$

$$aD_t^{\alpha} e = \frac{-1}{a}(\lambda - 1 + K_p)e + \frac{1}{a}\widetilde{W}\sigma(x) + \frac{1}{a}\widehat{W}\phi(e) + \frac{1}{a}K_i a D_t^{-\alpha} e - \frac{\gamma}{a}\left(\frac{1}{2} + \frac{1}{2}\widehat{W}^2 L_{\phi}^2\right) e \tag{23}$$

And if $w = \frac{1}{a} K_i a D_t^{-\alpha} e$, then $aD_t^{\alpha} w = \frac{1}{a} K_i e(t)$ [5]; then, we rewrite Eq. (23) as:

$$aD_t^{\alpha} e = \frac{-1}{a}(\lambda - 1 + K_p)e + \frac{1}{a}\widetilde{W}\sigma(x) + \frac{1}{a}\widehat{W}\phi(e) + w - \frac{\gamma}{a}\left(\frac{1}{2} + \frac{1}{2}\widehat{W}^2 L_{\phi}^2\right) e \tag{24}$$

We will show that the new state $(e, w)^T$ is asymptotically stable and the equilibrium point is $(e, w)^T = (0, 0)^T$, when $\widetilde{W}\sigma(x_r) = 0$, as an external disturbance.

Let V be, the next candidate Lyapunov function as [6]:

$$V = \frac{1}{2}\left(e^T, w^T\right)(e, w)^T + \frac{1}{2a} tr\left\{\widetilde{W}^T \widetilde{W}\right\} \tag{25}$$

The fractional order time derivative of Eq. (25) along with the trajectories of Eq. (24) is

$$aD_t^\alpha V = e^T aD_t^\alpha e + w^T aD_t^\alpha w + \frac{1}{a} tr\left\{aD_t^\alpha \widetilde{W}^T \widetilde{W}\right\} \tag{26}$$

$$\begin{aligned} aD_t^\alpha V = {} & e^T\left(\frac{-1}{a}(\lambda - 1 + K_p)e + \frac{1}{a}\widetilde{W}\sigma(x) + \frac{1}{a}\widehat{W}\phi(e) + w - \frac{\gamma}{a}\left(\frac{1}{2} + \frac{1}{2}\widehat{W}^2 L_\phi^2\right)e\right) \\ & + \frac{1}{a}\widetilde{W}^T K_i e + \frac{1}{a} tr\left\{aD_t^\alpha \widetilde{W}^T \widetilde{W}\right\} \end{aligned} \tag{27}$$

In this part, we select the next learning law from the neural network weights as in [7] and [8]:

$$tr\left\{aD_t^\alpha \widetilde{W}^T \widetilde{W}\right\} = -e^T \widetilde{W}\sigma(x) \tag{28}$$

Then, Eq. (27) is reduced to

$$aD_t^\alpha V = \frac{-1}{a}(\lambda - 1 + K_p)e^T e + \frac{e^T}{a}\widehat{W}\phi(e) + \left(1 + \frac{K_i}{a}\right)e^T w - \frac{\gamma}{a}\left(\frac{1}{2} + \frac{1}{2}\|\widehat{W}\|^2 L_\phi^2\right)e^T e \tag{29}$$

Next, lets consider the following inequality proved in [9]

$$X^T Y + Y^T X \leq X^T \Lambda X + Y^T \Lambda^{-1} Y \tag{30}$$

which holds for all matrices $X, Y \in \mathbb{R}^{nxk}$ and $\Lambda \in \mathbb{R}^{nxn}$ with $\Lambda = \Lambda^T > 0$. Applying (30) with $\Lambda = I$ to the term $\frac{e^T}{a}\widehat{W}\phi(e)$ from Eq. (29), where

$$e^T \widehat{W}\phi(e) \leq \frac{1}{2}\|\mathrm{e}\|^2 + \frac{1}{2}L_\phi^2\|\widehat{W}\|^2 \mathrm{e}^2 = \frac{1}{2}\left(1 + L_\phi^2\|\widehat{W}\|^2\right)\|\mathrm{e}\|^2$$

we get

$$aD_t^\alpha V \leq \frac{-1}{a}(\lambda - 1 + K_p)e^T e + \frac{1}{a}\left(\frac{e^T e}{2} + \frac{1}{2}\|\widehat{W}\|^2 L_\phi^2\right)e^T e + \left(1 + \frac{K_i}{a}\right)e^T w - \frac{\gamma}{a}\left(\frac{1}{2} + \frac{1}{2}\|\widehat{W}\|^2 L_\phi^2\right)e^T e \tag{31}$$

Here, we select $\left(1 + \frac{K_i}{a}\right) = 0$ and $K_v = K_i + 1$, with $K_v \geq 0$ then $K_i \geq -1$, with this selection of the parameters from Eq. (31) is reduced to:

$$aD_t^\alpha V \leq \frac{-1}{a}(\lambda - 1 + K_p)e^T e - \frac{(\gamma - 1)}{a}\left(\frac{1}{2} + \frac{1}{2}\|\widehat{W}\|^2 L_\phi^2\right)e^T e \tag{32}$$

Of the previous inequality, Eq. (32), we need that the fractional order Lyapunov derivative, $aD_t^{\alpha}V \leq 0$, to ensure that the trajectory tracking error is asymptotically stable, that is, $\lim_{t\to\infty} e(t) = 0$, which means that the nonlinear system follows the reference signal.

To achieve this purpose, we select:

$$\lambda - 1 + K_p > 0, a > 0, (\gamma - 1) > 0, aD_t^{\alpha}V \leq 0, \forall\, e,\ w,\ \widehat{W} \neq 0, e \neq 0$$

With the above Eq. (32), the control law that guarantees asymptotic stability of the tracking error is given by Eq. (33)

$$\begin{aligned} u = \Omega^{\dagger}[&-\widehat{W}\Gamma(z(x) - z(x_p)) - (A + I)(x - x_p) + K_p e + K_i aD_t^{-\alpha}e + K_v aD_t^{\alpha}e \\ &-\gamma\left(\frac{1}{2} + \frac{1}{2}\|\widehat{W}\|^2 L_{\phi}^2\right)e + f_r(x_r, u_r) - Ax_r - \widehat{W}\Gamma_z(x_r) - x_r + x_p \end{aligned} \tag{33}$$

Theorem: The control laws (Eq. (33)) and the adaptive weights (Eq. (28)) ensure that the trajectory tracking error between the fractional nonlinear system (Eq. (8)) and the fractional reference signal (Eq. (9)) satisfies $\lim_{t\to\infty} e(t) = 0$

Remark 2: *From Eq. (32), we have*

$aD_t^{\alpha}V \leq \frac{-1}{a}(\lambda - 1 + K_p)e^T e - \frac{(\gamma - 1)}{a}\left(\frac{1}{2} + \frac{1}{2}\|\widehat{W}\|^2 L_{\phi}^2\right)e^T e < 0, \forall\, e \neq 0,\ \forall\, \widehat{W}$, where V is decreasing and bounded from below by $V(0)$, and:

$V = \frac{1}{2}(e^T, w^T)(e, w)^T + \frac{1}{2a} tr\left\{\widetilde{W}^T \widetilde{W}\right\}$,; then we conclude that e, $\widetilde{W} \in L_1$; this means that the weights remain bounded.

5. Simulation

The manipulator used for simulation is a two revolute joined robot (planar elbow manipulator), as shown in **Figure 2**.

The dynamics of the robot is established by [10, 11], $M_{ij}(q),\ i,\ j = 1, 2$ of the inertia matrix $M(q)$ as

$$M_{11}(q) = m_1 l_{c1}^2 + m_2(l_1^2 + l_{c2}^2 + 2l_1 l_{c2}\cos(q_2)) + I_1 + I_2;$$

$$M_{12}(q) = m_2(l_{c2}^2 + l_1 l_{c2}\cos(q_2)) + I_2;$$

$$M_{21}(q) = m_2(l_{c2}^2 + l_1 l_{c2}\cos(q_2)) + I_2;$$

$$M_{22}(q) = m_2 l_{c2}^2 + I_2.$$

$$C_{11}(q, \dot{q}) = -m_2 l_1 l_{c2}\sin(q_2)\dot{q}_2;$$

$$C_{21}(q, \dot{q}) = -m_2 l_1 l_{c2}\sin(q_2)(\dot{q}_1 + \dot{q}_2);$$

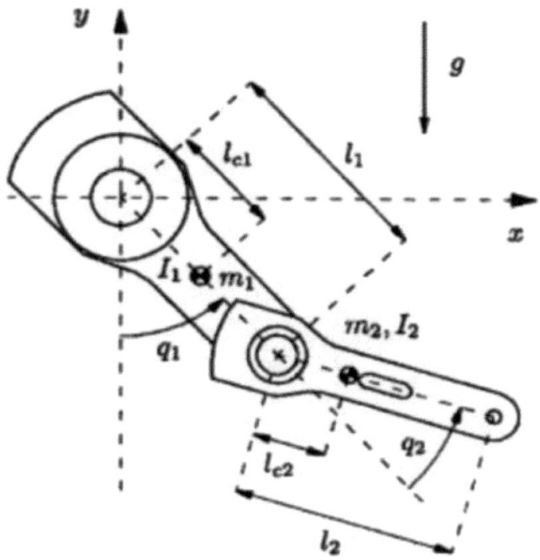

Figure 2. Diagram of the prototype planar robot with two degrees of freedom.

$$C_{21}(q,\dot{q}) = -m_2 l_1 l_{c2} \sin(q_2)\dot{q}_1$$

$$C_{22}(q,\dot{q}) = 0$$

And the torque vector $g(q)$:

$$g_1(q) = (m_1 l_{c1} + m_2 l_1) g sin(q_1) + m_2 l_{c2} g sin(q_1 + q_2);$$

$$g_2(q) = m_2 l_{c2} g sin(q_1 + q_2);$$

Thus, it is possible to write the equations of motion using the Lagrange equations for fractional manipulator system as [12]:

$$(m_1 + m_2) l_1^2 \ddot{\theta}_1 + m_2 l_1 l_2 \ddot{\theta}_2 \cos(\theta_2 - \theta_1) - m_2 l_1 l_2 \dot{\theta}_2^2 \sin(\theta_2 - \theta_1) + (m_1 + m_2) g l_1 \sin(\theta_1)$$
$$+ \frac{(\alpha - 1)}{(t - \tau)} \left[(m_1 + m_2) l_1^2 \dot{\theta}_1 + m_2 l_1 l_2 \dot{\theta}_2 cos(\theta_2 - \theta_1) \right] = Q_1$$

$$m_2 l_2^2 \ddot{\theta}_2 + m_2 l_1 l_2 \ddot{\theta}_1 \cos(\theta_2 - \theta_1) + m_2 l_1 l_2 \dot{\theta}_1^2 sin(\theta_2 - \theta_1) + m_2 g l_2 \sin(\theta_2)$$
$$+ \frac{(\alpha - 1)}{(t - \tau)} \left[m_2 l_2^2 \dot{\theta}_2 + m_2 l_1 l_2 \dot{\theta}_1 cos(\theta_2 - \theta_1) \right] = Q_2$$

The terms containing α indicate the additional terms resulting from the fractional order model and the right-hand sides denote the generalized force terms resulting from the forcing functions, and there is a specific set of values for Q_1 and Q_2 for each case.

With the end of supporting the effectiveness of the proposed controller, we have used a Duffing equation.

The fractional order neural network is modelling by the differential equation:

$aD_t^\alpha x_p = \mathrm{A}(x) + \mathrm{W}^*\Gamma_z(x) + \Omega_u$, with $A = -\lambda I, I \in \mathbb{R}^{4x4}$, and $\lambda = 20$, W^* is estimated using the learning law given in Eq. (28).

$\Gamma_z(x) = (\tanh(x_1), \tanh(x_2), \ldots, \tanh(x_n))^{\mathrm{T}}, \Omega = \begin{pmatrix} 0 & 0 & 1 & 0 \\ 0 & 0 & 0 & 1 \end{pmatrix}^{\mathrm{T}}$ and the u is calculated using Eq. (33). The plant is stated in [3] and [13], and it is given by:

$$D(q)\ddot{q} + C(q,\dot{q})\dot{q} + G(q) = \tau$$

We try to force this manipulator to track a reference signal [14] given by undamped Duffing equation:

$$\ddot{x} - x + x^3 = 0.114\cos(1.1t) : x(0) = 1, \dot{x}(0) = 0.114$$

To get the fractional order Duffing's system, this equation can be rewritten as a system of the first-order autonomous differential equations in the form [15]:

$$\frac{x(t)}{dt} = y(t)$$

$$\frac{y(t)}{dt} = x(t) - x^3(t) - \alpha y(t) + \delta\cos(\omega t)$$

Here, the conventional derivatives are replaced by the fractional derivatives as follows:

$$aD_t^\alpha x(t) = y(t)$$

$$aD_t^\alpha x(t) = x(t) - x^3(t) - \alpha y(t) + \delta\cos(\omega t)$$

where α is the fractional orders and α, δ, ω are the system parameters.

Illustrated, the response in the time, angular position and torque applied to the fractional nonlinear system are shown in **Figures 3–7**. As can be observed, the trajectory tracking objective is obtained

$$\alpha = 1, \quad \beta = 1$$

Its phase space trajectory is given in **Figure 8**, and the time evolution for the position angles and applied torque are shown in **Figures 9–12**. As can be seen in **Figures 9** and **10**, the trajectory tracking is successfully obtained where plant and reference signals are the same.

$$\alpha = 0.99, \quad \beta = 0.99$$

http://dx.doi.org/10.5772/intechopen.70020

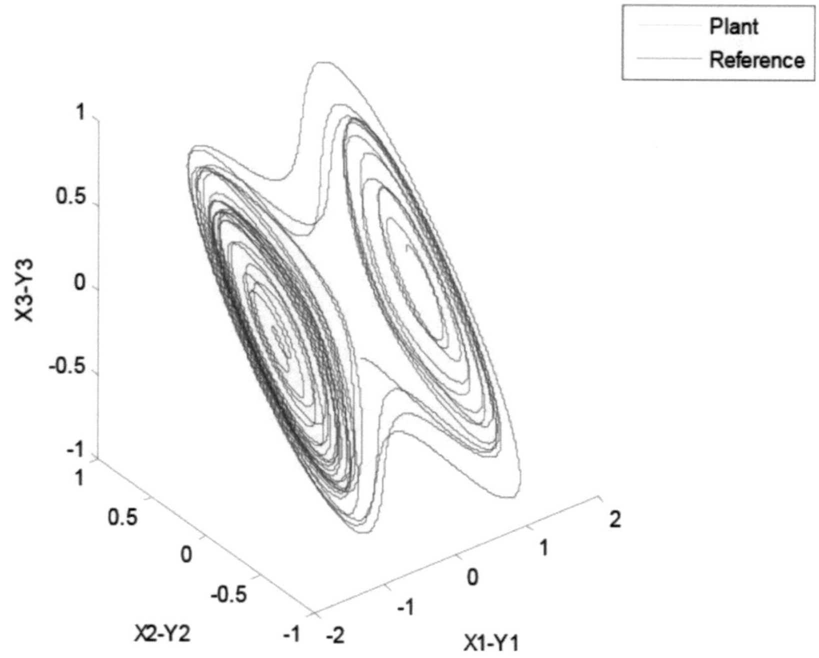

Figure 3. A phase space trajectory of Duffing equation.

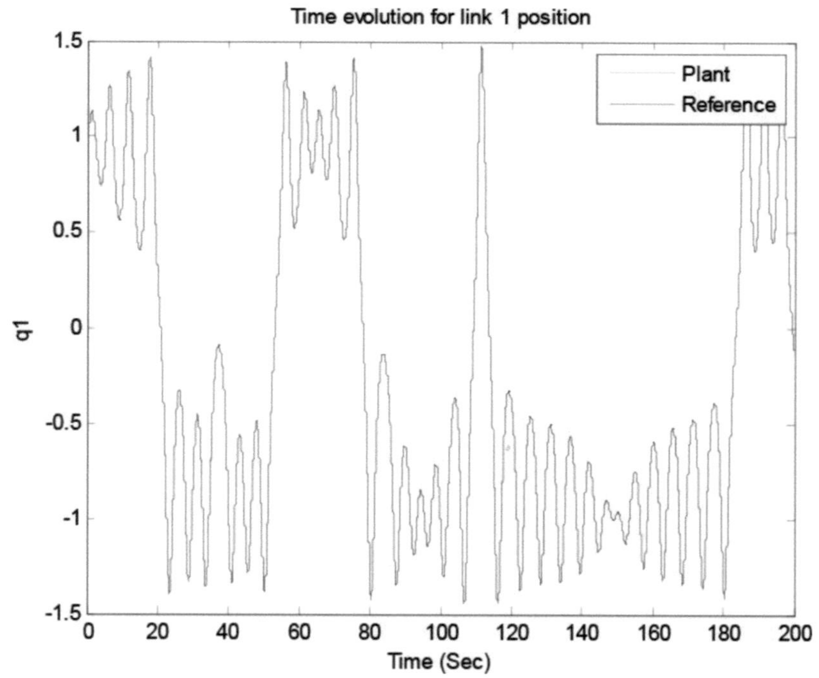

Figure 4. Time evolution for the angular position q_1 (rad) of link 1.

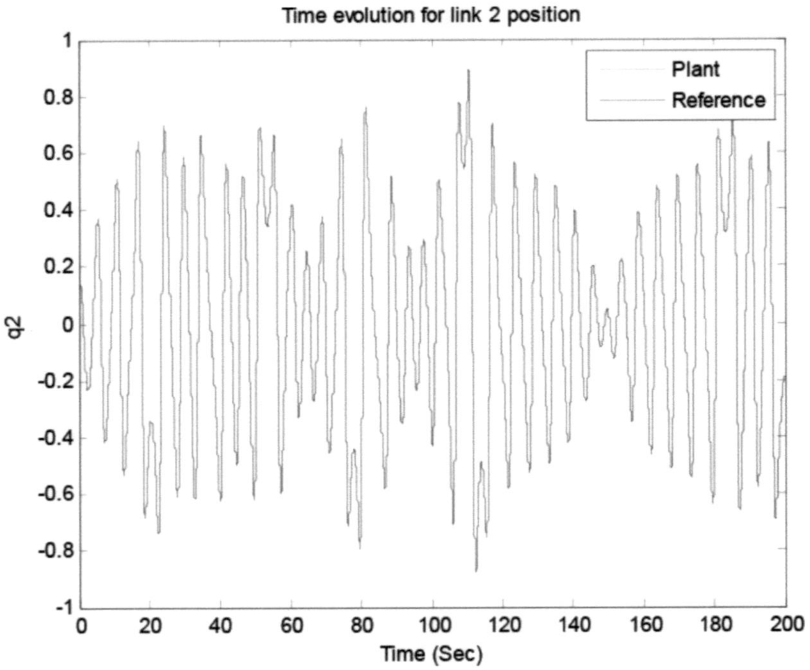

Figure 5. Time evolution for the angular position q_2 (rad) of link 2.

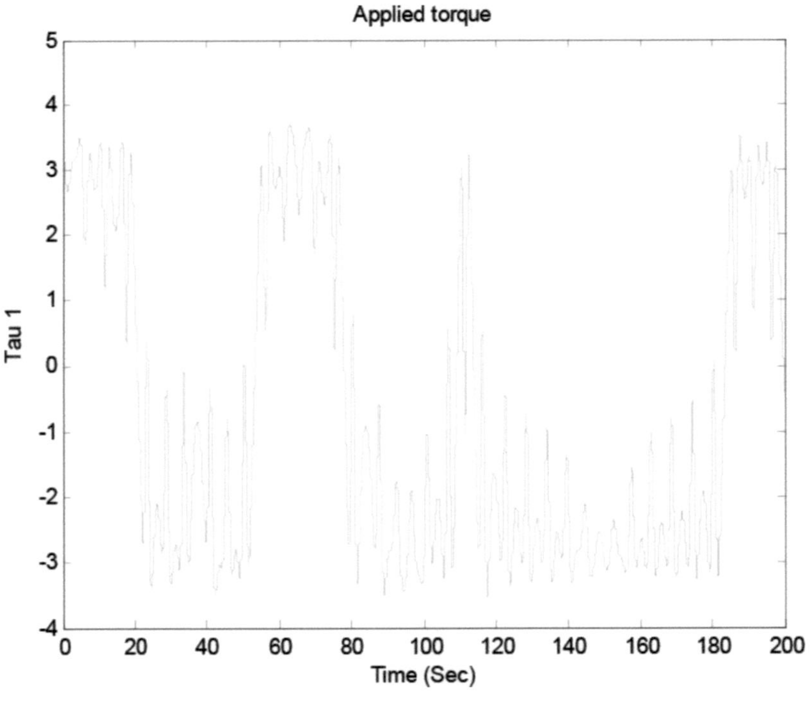

Figure 6. Torque (Nm) applied to link 1.

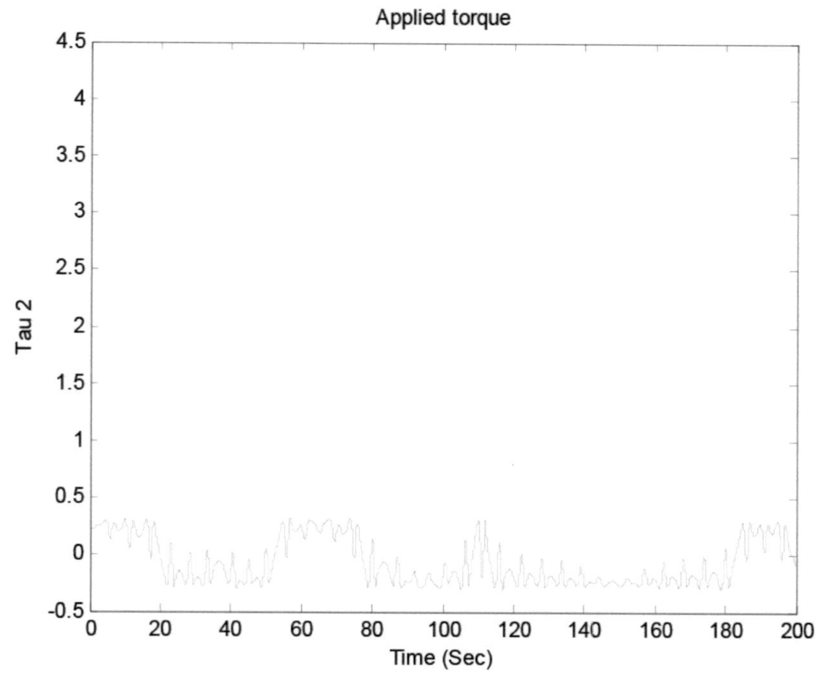

Figure 7. Torque (Nm) applied to link 2.

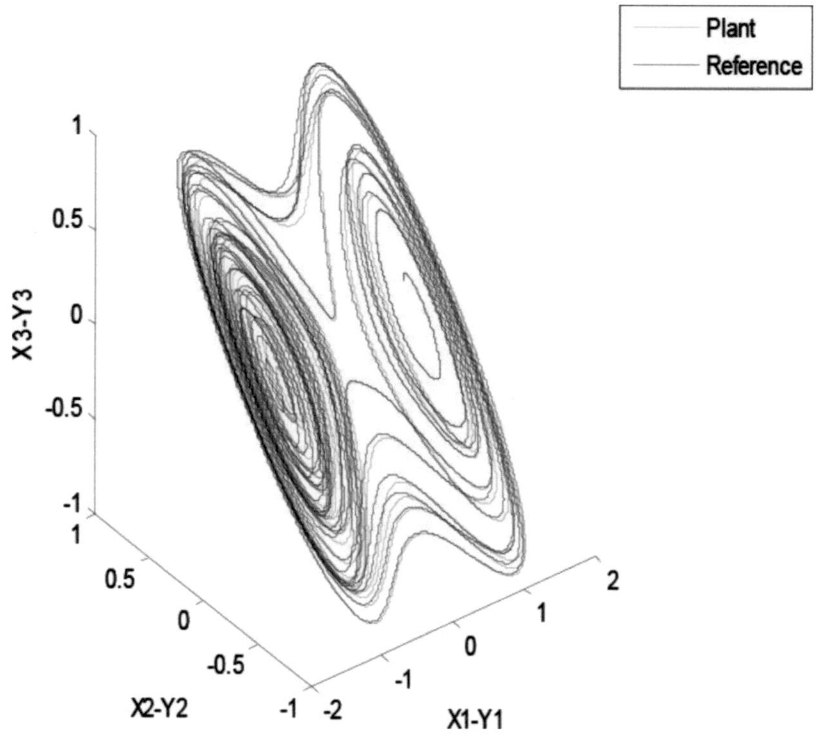

Figure 8. A phase space trajectory of Duffing equation.

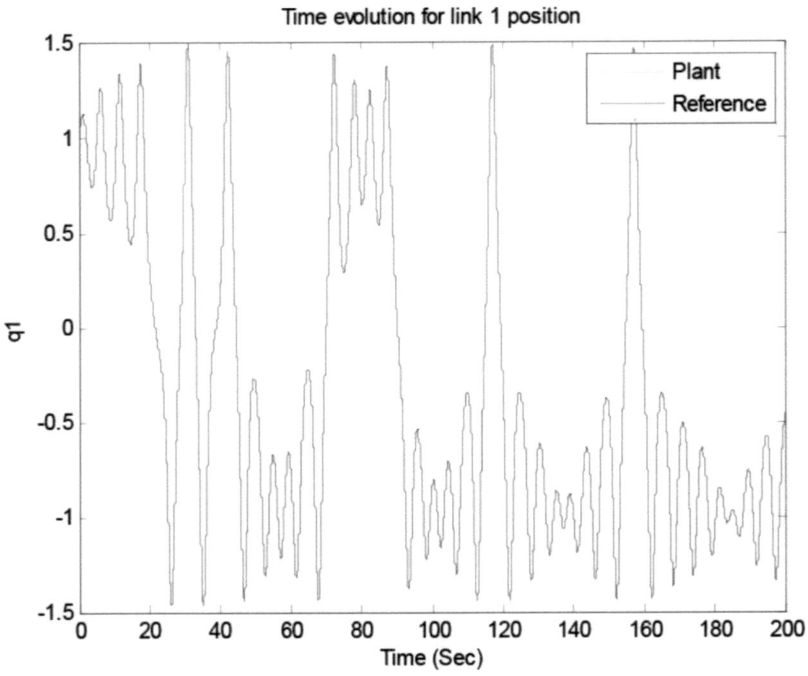

Figure 9. Time evolution for the angular position q_1 (rad) of link 1.

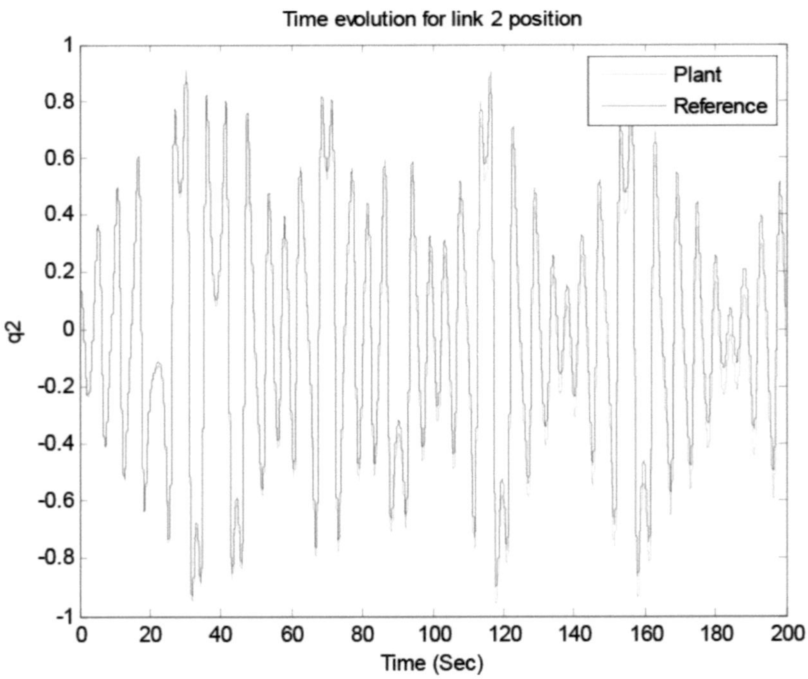

Figure 10. Time evolution for the angular position q_2 (rad) of link 2.

http://dx.doi.org/10.5772/intechopen.70020

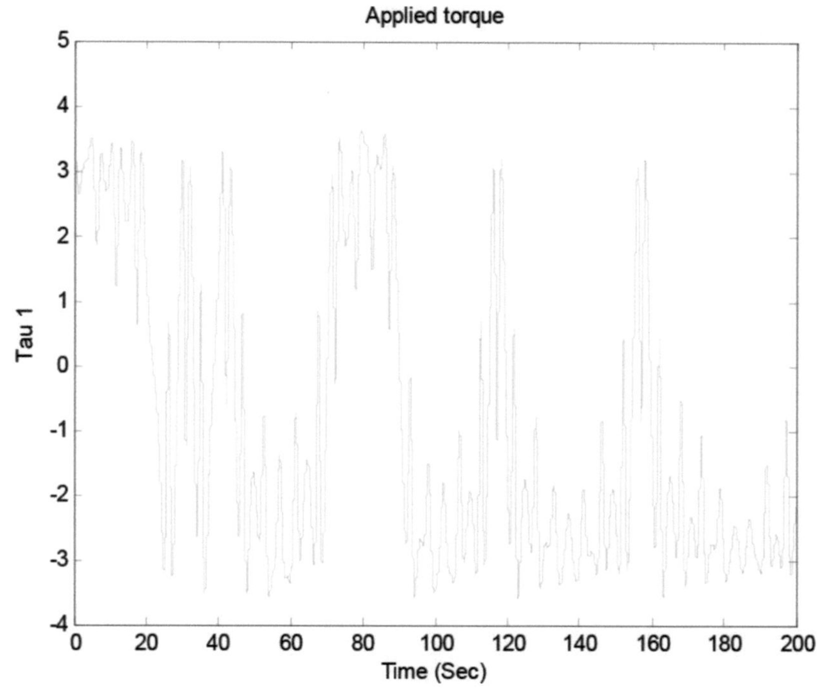

Figure 11. Torque (Nm) applied to link 1.

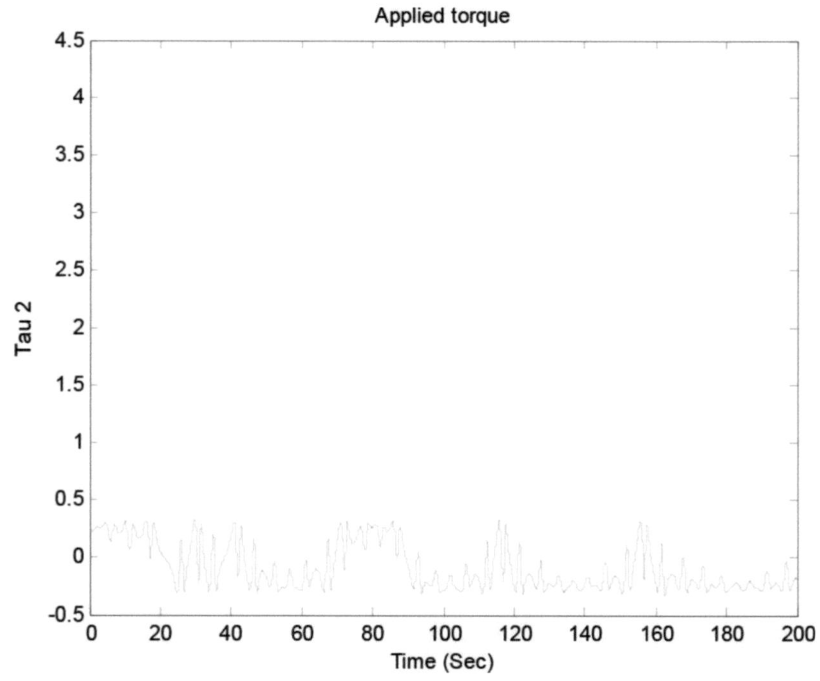

Figure 12. Torque (Nm) applied to link 2.

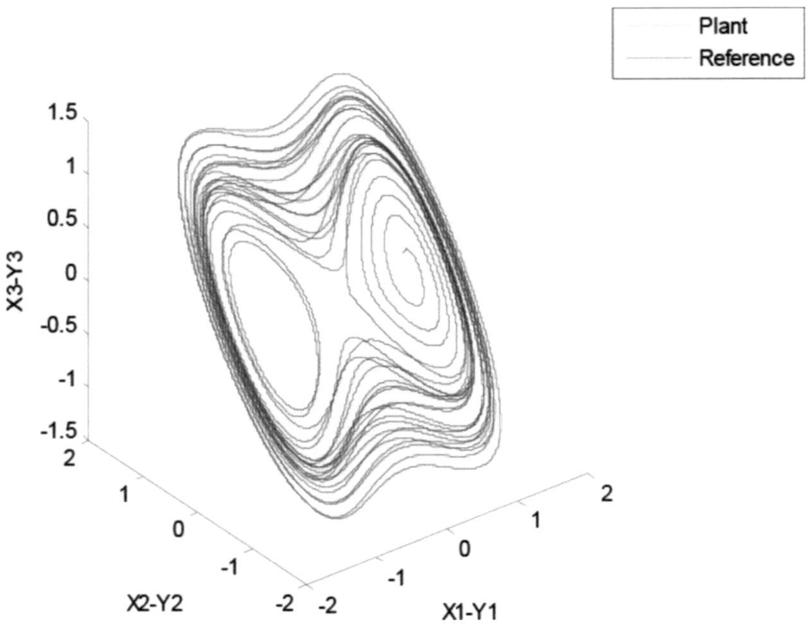

Figure 13. A phase space trajectory of Duffing equation.

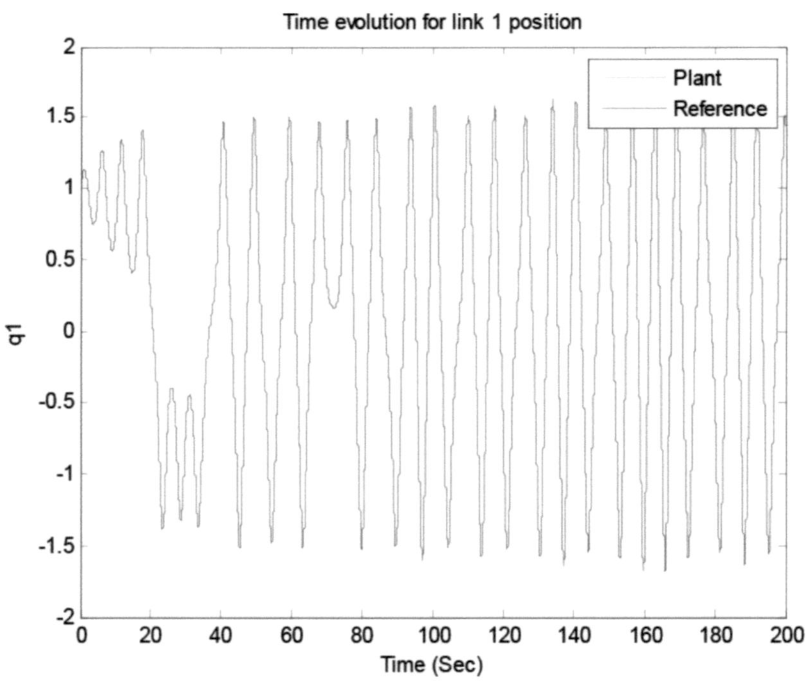

Figure 14. Time evolution for the angular position q_1 (rad) of link 1.

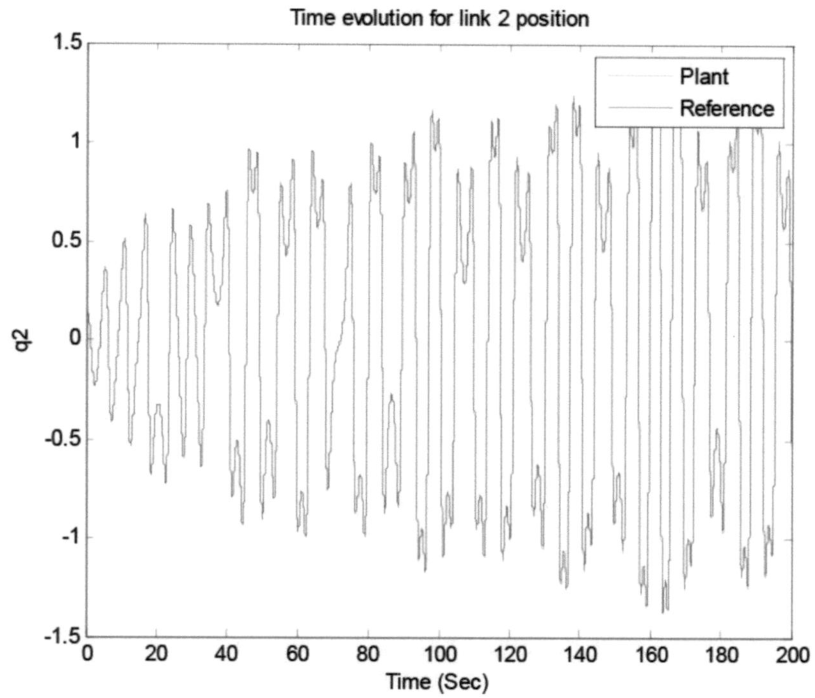

Figure 15. Time evolution for the angular position q_2 (rad) of link 2.

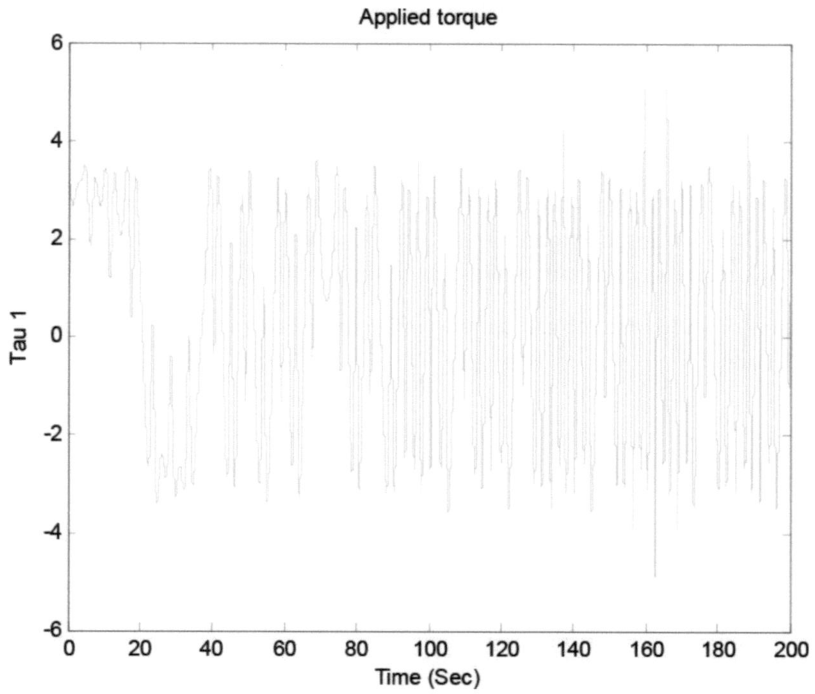

Figure 16. Torque (Nm) applied to link 1.

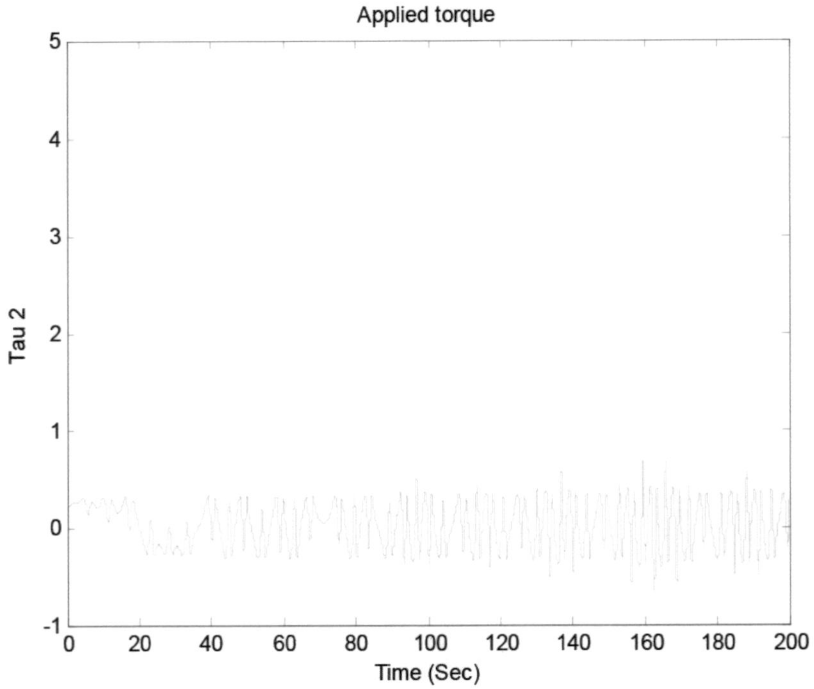

Figure 17. Torque (Nm) applied to link 2.

Its phase space trajectory is given in **Figure 13**, and the time evolution for the position angles and applied torque are shown in **Figures 14–17**. As can be seen in **Figures 14** and **15**, the trajectory tracking is successfully obtained where plant and reference signals are the same.

$$\alpha = 0.001, \quad \beta = 0.001$$

As can be observed, in the graphs of the trajectory tracking, the experimental results obtained in this chapter show a good experimental performance. The laws of control are obtained online, as well as the laws of adaptive weights in the fractional order neural network.

The control laws obtained are robust to modeling errors and nonmodeled dynamics (unknown nonlinear systems).

6. Conclusions

We have discussed the application of the stability analysis by Lyapunov of fractional order to follow trajectories of nonlinear systems whose mathematical model is unknown. The convergence of the tracking error is established by means of a Lyapunov function, as well as a control law based on Lyapunov and laws of adaptive weights of fractional order dynamical neural networks.

The results show a satisfactory performance of the fractional order dynamical neural network with online learning.

Acknowledgements

Authors thank Mexican National Science and Technology Council, (CONACyT), Mexico.

The first and second authors thank the Autonomous University of Nuevo Leon, Dynamical Systems Group of the Department of Physical and Mathematical Sciences (FCFM-UANL), Mexico.

The third author thanks the University of Monterrey (UDEM), Mexico.

Author details

Joel Perez P.[1]*, Jose Paz Perez P.[1] and Martha S. Lopez de la Fuente[2]

*Address all correspondence to: joel.perezpd@uanl.edu.mx

1 Department of Physical and Mathematical Sciences, Autonomous University of Nuevo Leon (FCFM, UANL), San Nicolas de los Garza, Nuevo Leon, Mexico

2 University of Monterrey (UDEM), Mexico

3 Department of Electrical Engineering, Autonomous University of Nuevo Leon (FIME, UANL), San Nicolas de los Garza, Nuevo Leon, Mexico

4 Bilingual Progressive Bachelor, No. 16, Autonomous University of Nuevo Leon (Prepa 16, UANL), San Nicolas de los Garza, Nuevo Leon, Mexico

5 Department of Physical and Mathematical Sciences, Autonomous University of Nuevo Leon (UANL), San Nicolas de los Garza, Nuevo Leon, Mexico

References

[1] Zhang S, Yu Y, Wang Q. Stability analysis of fractional-order Hopfield neural networks with discontinuous activation functions. Neurocomputing. 2016;**171**:1075–1084. Available from: www.elsevier.com/locate/neucom

[2] Joshi SD, Talange DB. Integer & fractional order PID controller for fractional order subsystems of AUV. In: 2013 IEEE Symposium on Industrial Electronics & Applications (ISIEA 2013); 22–25 September 2013; Kuching, Malaysia

[3] Kelly R, Haber RE, Haber-Guerra RE, Reyes F. Lyapunov stable control of robot manipulators: A fuzzy self-tuning procedure. Intelligent Automation and Soft Computing. 1999;**5**(4):313–326

[4] Moreno AR, Sandoval VJ. Fractional order PD and PID position control of an angular manipulator of 3DOF

[5] Grigoletto EC, de Oliveira EC. Fractional versions of the fundamental theorem of calculus. Applied Mathematics. 2013;**4**:23-33. DOI: 10.4236/am.2013.47A006. Available from: http://www.scirp.org/journal/am

[6] Chen D, Zhang R, Liu X, Ma X. Fractional order Lyapunov stability theorem and its applications in synchronization of complex dynamical networks. Communications in Nonlinear Science and Numerical Simulation. 2014;**19**:4105-4121. Available from: www.elsevier.com/locate/cnsns

[7] Rovitahkis GA, Christodoulou MA. Adaptive Control with Recurrent High-Order Neural Networks. New York, USA: Springer Verlag; 2000

[8] Ioannou PA, Sun J. Robust Adaptive Control. Upper Saddle River, NJ: PTR Prentice-Hall. ISBN: 0-13-439100-4

[9] Ma W, Li C, Wu Y, Wu Y. Adaptive synchronization of fractional neural networks with unknown parameters and time delays. Entropy. 2014;**16**:6286-6299. DOI: 10.3390/e16126286. ISSN: 1099–4300

[10] Reyes F, Kelly R. Experimental evaluation of model-based controllers on a direct-drive robot arm. Mechatronics. 2001;**11**:267–282

[11] Meza JL, Santibáñez V, Soto R, Perez J, Perez J. Analysis via passivity theory of a class of nonlinear PID global regulators for robot manipulators. In: Advances in PID Control. 2011. ISBN: 978-953-307-267-8

[12] David SA, Balthazar JM, Julio BHS, Oliveira C. The fractional-nonlinear robotic manipulator: Modeling and dynamic simulations. In: 9th International Conference on Mathematical Problems in Engineering, Aerospace and Sciences. AIP Conference Proceedings. Vol 1493. American Institute of Physics; 2012. pp. 298-305. DOI: 10.1063/1.4765504. ISBN: 978-0-7354-1105-0

[13] Meza JL, Santibañez V, Soto R, Llama MA. Stable fuzzy self-tuning PID control of robot manipulators. In: Proceedings of the 2009 IEEE International Conference on Systems, Man, and Cybernetics; October 2009; San Antonio, TX, USA

[14] Li Z, Chen D, Zhu J, Liu Y. Nonlinear dynamics of fractional order Duffing system. Chaos, Solitons and Fractals: The Interdisciplinary Journal of Nonlinear Science, and Nonequilibrium and Complex Phenomena. 2015;**81**(Part A):111-116. DOI: 10.1016/j.chaos.2015.09.012

[15] Petras I. Fractional-Order Nonlinear Systems: Modeling, Analysis and Simulation. Nonlinear Physical Science. Heidelberg, Dordrecht, London, New York: Springer. ISBN: 978-3-642-18100-9. Series ISSN: 1867–8440

Section 3

Legal Impacts

Chapter 4

Robots Liability: A Use Case and a Potential Solution

Alejandro Zornoza, José C. Moreno,
José L. Guzmán, Francisco Rodríguez and
Julián Sánchez-Hermosilla

Additional information is available at the end of the chapter

http://dx.doi.org/10.5772/intechopen.69888

Abstract

In this chapter, a system of distribution of responsibility for damages caused by robots is introduced, and its practical application on the results obtained in the real experiences at the University of Almería (Spain) is analyzed. The attribution of liability for damages produced by autonomous agents usually focuses the theoretical discussion on legal and ethical fields on robotics. The European Parliament adopted the report with recommendations to the Commission on Civil Law Rules on Robotics (2015/2103(INL)) in February 2017. This work includes the master guidelines that the European Commission should take into account to legislate this technology. In its attempt to attribute responsibility for damages caused by robots, the Committee considers that once responsible parties have been identified, their liability level should range, looking the robot's learning capability and the knowledge learned from its owner. This work proposes the use of responsibility setting matrix as a mechanism to distribute liabilities between the robot, the manufacturer, and the owner, depending on the knowledge programmed by the manufacturer and the one acquired by the robot (through its learning ability and the adjustments made by the owner), that would distribute the responsibility for damages among the three agents involved.

Keywords: Fitorobot, Robotic Liability Matrix, robots, liability, responsibility

1. Introduction

One of the main current issues about robotics is its liability regimes, which is especially referred to robots with a high grade of autonomy.

Assuming that the term "responsibility" is a cultural product (it is the result of the interaction between sociability and moral agency; hence, each state has his own liability rules), we are going to introduce the technology as objective element for rethinking the responsibility.

The first works of the European Union (EU) on this subject call for new rules which focus on how a machine can be held responsible for its acts or omissions, suggesting the possibility to create a new legal status only for robots.

This approach would have important social, political, legal, and economical implications that we are not going to expose in this work. On the one hand, we consider robots as tools. Without undertaking any ethical and philosophical evaluation, robots are created to do and to make human tasks. If the labor is physical (e.g., to carry weight) or is psychological (e.g., to perform complex mathematical operations), it is not relevant; they are products and will be considered as such. There is a tendency to humanize robots because they have "human behavior." This is a consequence of their artificial intelligence and the type of work they do, but the most important part of that term is "artificial"; this means that they are something manufactured by human being and could be controlled, limited, and, so, equated to any other product.

On the other hand, there are different kinds of robots, so there should be different types of responsibility. Some robots have physical structure, as care robots or social robots, and some robots only exist on a digital form, as trader bots. When it concerns damage, one usually thinks directly of a robot that hurts a person, but observing the different kinds of robots, the damage can be classified as physical (over persons or goods), economic, or spiritual (moral damage). In order to guide the discussion, it should be noticed that we have observed the state of the art to write these lines, so we have employed a real use case.

In this chapter, Fitorobot is introduced as case study. Further on, the existing product liability regimes will be explained and applied to robots. This issue will allow to show the EU's agreed targets and to expose the adequacy or insufficiency of the current law for it. Finally, a genuine proposal to rework the responsibility and the way to apply it will be described.

2. Fitorobot: a use case

Tracked mobile robots (TMR) are increasingly being used on rough off-road terrains for applications such as forestry, mining, agriculture, and army, and in general in many kinds of applications on unpaved terrains. These applications usually require robots to travel across unprepared terrains performing some activity or transporting materials.

In this chapter, we use a research project at the University of Almería (Spain), which is devoted to the development of an TMR to perform different tasks in greenhouses (especially those related with spraying activities), as a use case.

This TMR is called *Fitorobot* (see **Figure 1**); it is a robot with a mass of 756 kg (with the spray tank full), and the dimension is 1.5 m long × 0.7 m wide. It is driven by a powerful 20 HP gasoline engine. It is equipped with several sensors, but in the real tests carried out in this work, only four have been used (right track encoder, left track encoder, radar, and magnetic compass). The real track radius of the test bed is 0.15 m, but the calibrated track radius is 0.10 m. The distance

Figure 1. Fitorobot [1].

between the track centers is 0.5 m. More details about the features of this TMR can be found in Ref. [1].

Autonomous navigation of the mobile robot relies heavily on external sensor information. Therefore, the performance of the robot navigation will depend heavily on the installed sensors on the platform. Therefore, different types of sensors are installed on the platform, where some of them are redundant for the purpose of testing different configurations and for security reasons.

Table 1 summarizes the sensors installed on the Fitorobot. One middle-range sonar and four short-range sonars have been located in front and in each side of the platform, respectively. These sensors enable the robot to sense the environment and the greenhouse corridors. Odometry establishes the position and velocity of the robot, using two incremental optical encoders attached to the axle of rotation of the track motors. One radar and one magnetic compass measure the linear velocity and orientation of the vehicle, respectively. It is protected from unexpected obstacles in the environment by a security sensor composed of four tactile bars around the vehicle. Finally, a pressure sensor has been installed in the spraying hydraulic system to regulate the spraying controllers [1]. The sensor positions on the platform are shown in **Figure 2**.

Sensor	Mark, model	Range
Pressure	WIKA, ECO-1	0–250 bar
Sonar (middle dist.)	Siemens, Bero III	40–300 cm
Sonar (short dist.)	Siemens, Bero M18	15–100 cm
Magnetic compass	KVH, C100	0–360°
Radar	LH Agro, Compact II	0.08–17.3 m/s
Encoders	Sick, DRS61	0–360°
Security sensor	SafeWork, SKL25-40	On-off
Camera	Logitech 2 Mpixel QuickCam Sphere AF webcam	1600 × 1200

Table 1. Features of the installed sensors.

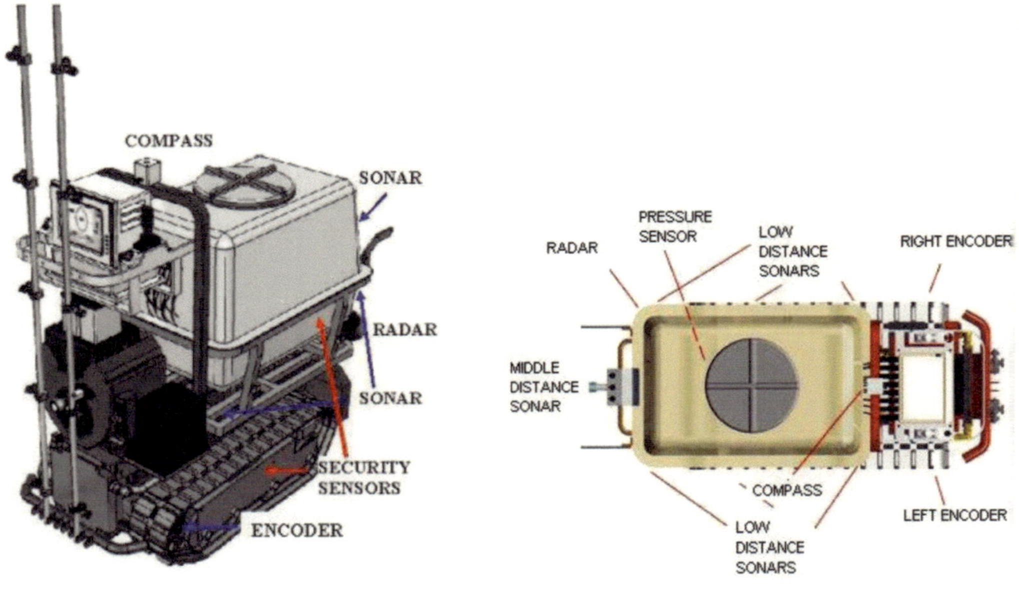

(a) Lateral view of the sensorial system

(b) Top view of the sensorial system

Figure 2. Sensors on the experimental platform [1].

A more detailed information about the actuator and control system of the Fitorobot can be found in Ref. [1].

3. Degrees of freedom for influencing the robot behavior

The behavior of a robot is given by the set of algorithms that constitute its artificial intelligence. These algorithms must allow the robot to develop the set of tasks for which it has been designed, respecting the fundamental laws of robotics [2]:

1. A robot may not injure a human being or, through inaction, allow a human being to come to harm.
2. A robot must obey the orders given by human beings except where such orders would conflict with the first law.
3. A robot must protect its own existence as long as such protection does not conflict with the first or second laws.

In Ref. [3], a revised set of five laws has been proposed:

1. Robots are multi-use tools. Robots should not be designed solely or primarily to kill or harm humans, except in the interests of national security.
2. Humans, not robots, are responsible agents. Robots should be designed and operated as far as practicable to comply with existing laws, fundamental rights, and freedoms, including privacy.
3. Robots are products. They should be designed using processes which assure their safety and security.
4. Robots are manufactured artifacts. They should not be designed in a deceptive way to exploit vulnerable users; instead, their machine nature should be transparent.
5. The person with legal responsibility for a robot should be attributed.

In order to fulfill the fifth law, some considerations about the possibilities to modify the robot's predefined behavior must be taken into account. So, in this chapter we consider five levels of behavior modification, from a closed architecture to a completely open architecture:

a. Closed architecture

In this category we include all robots with a non-customizable predefined behavior. Examples of this type of robots are the household robots, as Roomba by iRobot, and entertainment and leisure robots, as Robotic Balls by Sphero. A closed version for Fitorobot can be considered. So, the sole responsibility of the farmer is to use the robot properly.

b. Autotuning from user behavior

This category includes robots that learn from the user. The driverless car is one of the clearest example of mobile robots. With the last advances in this technology, it is possible that in the near future these cars, in order to satisfy the particular user expectations, incorporate an algorithm that learns from the user skills and his driving styles.

c. Setup parameter calibration

Algorithms controlling the robot behavior in general have parameters depending on the context or particular application of the robot. So, for example, if an artificial vision algorithm is used to navigate, there will be some parameters for the algorithm that will depend on the characteristic context for the environment navigated by the robot. For example, if we consider the Fitorobot presented in Section 2, we need to fix the umbral value for flooring color so that the robot can determine the way to follow centering itself between plants. **Figure 3** shows two different types of flooring, a soil cropping greenhouse and an hydroponic crop greenhouse with mulching. For both cases different umbralization values must be chosen in a manual way, so if the incorrect value is chosen by the farmer, the greenhouse corridors will not be correctly identified, and the Fitorobot will collide with the plants causing a disaster with important economic consequences. **Figure 4** shows an example of segmented corridor.

(a) Soil cropping greenhouse (b) Hydroponic crop greenhouse with mulching

Figure 3. Different types of a greenhouses flooring.

Figure 4. An example of a segmented corridor marking the trajectory to follow by the Fitorobot.

Two types of calibration may be considered:

i. Automatic calibration: The algorithm automatically sets the parameters. In this case, the robot identifies the most adequate values for the particular environment. However, a problem may occur if the automatic calibration run under conditions other than those used in the calibration stage.

ii. Manual calibration: Parameters are set by the user. In this case, the user chooses the value for parameters, so the responsibility for the good performance of robot lies with the user.

d. High-level programming

There are robots that can be programmed but with certain limitations. So, the maker partially opens the architecture robot, so that the user can customize its behavior in a single way. An example of robot included in this category is Aisoy, a social robot by Aisoy Robotics, in which speaker, microphone, camera, contact sensors, and single movements of the head, eyelids, eyebrows, and body can be programmed in order to interact with a person.

e. Low-level programming

This is the most customizable category. All hardware and software elements can be modified in order to obtain a determined behavior. This category would include the research robots. However, other end-user robots may be also considered, although it is important to note that these robots will commonly be programmed by an intermediate agent between the maker, the user, and the engineer. From the user point of view, the robot would have a closed architecture. Fitorobot is also an example of this type of robot.

4. Robots liability

Action and responsibility are two close concepts; we are responsible for all of our actions because our behavior is guided by reasons. Something similar occurs with robots.

The Committee on Legal Affairs of the European Union considers that the more autonomous robots are, the less they can be considered simple tools in the hands of manufacturers, owners, or users. As a consequence, the ordinary rules on liability are insufficient.

To fully manage this issue, two work lines are proposed. The first one is to establish a direct link between the robot behavior and the knowledge learned from the owner. In this sense, the owner plays a teacher role, so the longer a robot's *education* has lasted, the greater the responsibility of its owner should be.

The second one is the possibility to create a specific legal status for the most sophisticated robots (named *electronic persons*) and to apply electronic personality to cases where robots make smart autonomous decisions or otherwise interact with the third parties. That last argument has an abstract, legal, and ethical meaning that is too different from lines of this work, so we are going to focus on the first point.

We will begin by saying that although the Committee highlights that robots need special rules based on their nature, robots are not so different from other industrial and smart mechanical tools. Robots are machines, and all machines can produce damages because they are badly handled or because they are defecting products.

One of the most serious problems is the European legal analysis of defect concept, in terms of finding similarities or contrast between different court points of view. The differences of perspective and of process will have fundamental importance in the practical decision-making [4].

A summary of legal aspects about liability is presented in the following paragraphs. Considering robots as products, we are going to introduce different liability regimes depending on the subject, focusing on cases of shared liability for making a debate more interesting than the strict responsibility of the manufacturer or the user.

4.1. The liability of the producer

The manufacturer (or producer, as named Directive 85/374/EEC of 25 July 1985 on the approximation of the laws, regulations, and administrative provisions of the member states concerning liability for defective products) can mean:

1. The producer of a raw material and the manufacturer of a finished product or of a component part
2. The importer of the product
3. Any person putting their name, trademark, or other distinguishing features on the product
4. Any person supplying a product whose producer or importer cannot be identified

This multi-identity situation is a consequence of *pro consumatore* doctrine: an unknown producer cannot be a reason for unprotecting the consumer for damage caused by his products.

From a general point of view, we can say that a robot is composed of two things: software and hardware. In effect, it should do the distinction between producer and programmer, being the first person responsible for the whole electronic (sensors and actuators) and mechanical parts and the second person responsible for the intern process of the robot (learning capability, image processing, decision-making process, etc.).

Currently, and observing the lines of responsibility drawing by Directive 85/374/EEC, the difference between both is not relevant for the injured person, but in order to propose a new model and considering that the user can be the programmer sometimes, the labeling of the Directive could not be enough.

The producer can be responsible on two ways: based on detrimental qualities of his products (defective product) and based on the product behavior (negligence).

A product is defective when it does not provide the safety which a person is entitled to expect, taking all circumstances into account, including:

1. The presentation of the product.
2. The use to which the product could reasonably be expected to serve; in the legal practice, this also involves designing the product to avoid inappropriate use of situations. In consequence, it is not possible to avoid the defect product category just complying with the existing European and national regulations or requirements (e.g., ISO rules). In the field of

robotics, this consideration can be substantial, even if we talk about limiting the autonomy of robots as a way to prevent damages.

3. The time when the product was put into circulation.

The presentation of the product is not advertising and packaging only; it is the user manual and the information about components, characteristics, mode of use, and contraindications too. The presentation must be clear and detailed.

The injured consumer must prove that the damage is actual and caused by a defect in the product, but he/she does not have to prove the negligence or fault of the producer or importer; a causal link between the damage and the defect is enough.

Following the law, there are three categories of deficiencies: first, the product does not correspond to those of its same series (manufacturing defect). The second one consists of products that have a fault in its conception, so the deficiencies affect the whole series (design fault). And, third, the products have a perfect design and manufacturing but could be potentially dangerous if the user does not have adequate information to use it (defect of information).

In this respect, the Council of European Convention in 1977 rejected the adoption of an exclusive liability regime for dangerous products. This criterion has been followed later in the regulations on this matter. The dangerous product does not mean unsafe product: the safety is in relation to the handling and consequences arising from its use. A product can be dangerous and safe at the same time.

Observing this classification, we do not consider that a new category of responsibility for damages produced by a robot is necessary. Robots will be as unsafe as producers want to make them.

4.2. The liability of the programmer

We must distinguish two kinds of programmers: the programmer in the proper sense, whose job is to prepare the robot for be used by the customer, and the user-programmer, who is a user that programs the robot with the limitations established by the manufacturer.

4.2.1. The programmer

The programmer will be responsible for damages caused by robots when the damage is related to software failures, faults or errors.

A *failure* is an event that occurs when the correct service deviates, because it does not comply with the functional specification or because this specification did not adequately describe the system function. The *error* is defined as part of a system's total state that may lead to a failure. And, the cause of the error is called *fault* [5].

If a user acquired a closed architecture robot, any damage produced by the machine is the liability of programmer and so a product liability case.

The producer assumes to own all acts and omissions by his/her employees which occur during the course of their employments. If the programmer is employed by the manufacturer, the assumption of liability for one or the other is significant for his domestic political only. And, if the programmer is external, the producer assumes his/her acts too, because someone contributes to produce the final results but it has a procedural advantage: the producer can direct an action of compensation for damages to the external programmer.

4.2.2. The user-programmer

When the robot allows a degree of personalization, there is an effect of displacement of liability and voluntary assumptions of risk by the user.

This displacement cannot be understood as an absolute release of producer responsibility; the starting point in this regard should be the assumption that the user is not an expert on robotics and he/she only uses the technology that others create. Indeed, the assumption of risk by the user is limited to the risk that he/she may know. The duty of information about the conditions and mode of use has an essential role to play in this case.

The customizable robots are not "empty"; they have a minimum knowledge on which to work, given by the factory software. The distinction must be made between wrong customization-calibration and wrong basis programming. In the case of damage caused by a customized robot, the user would be responsible if the injury has originated in wrong final programming only. However, if the damage had occurred in any way, regardless of whether the robot was well or badly personalized because it is a factory of software problem, the user would not be responsible.

Taking a risk only becomes negligence if the conduct in risk-taking is unreasonable. Negligence is the interference with the duty to take care, causing damage or injury to another person.

4.3. The liability of the owner

By Directive 2011/38/EU of the European Parliament and of the Council of October 25 2011 on consumer rights, *consumer* means any natural person who is acting for purposes which are outside his trade, business, craft, or profession, and *trader* means any natural person or any legal person who is acting for purposes relating to his trade, business, craft, or profession.

Literally, the farmer who trades with his/her crops and acquires Fitorobot should not be covered by Consumer Law because the robot is intended for professional use. Nevertheless, there is an exception: tort rules are applied to all injured people, and as such the direct purchaser, the holder, the bystander, or the professional are identified.

Accordingly, the trader farmer who got Fitorobot is covered by some points of Consumer Law if he/she is injured by the robot, and there is no blame or negligence in his/her conduct, acting the preceding regimes, depending on the case.

To speak about the liability of the owner, a distinction is required: we must handle rightly the concepts of owner and user. The owner is the person that has purchased the product. An owner has duties which include machine's maintenance, preservation, and upgrading. The user is the person who can use the product. Both can be the same person or not.

In the case of Fitorobot acquired by a trader farmer, the farmer is the owner. If he/she has an employee who has permission to use Fitorobot, the employee is the user. In this case, the regime previously explained about dependence between producer and programmer is applied, and if a third party suffers a damage caused by the user's fault, he/she can act against the owner, and the owner can repeat the action against the user later. For this reason, we are going to do reference to *owner* always, because in our case he/she is the ultimate responsible. The owner will be responsible for negligence.

Negligence modulates the liability regimes of other subjects. The legal regime of product liability is imperative. That means that it cannot be modulated by the will of the parties, except if the doctrine *pro consumatore* is applied: any alteration of the manufacturer's liability should be for aggravating it. So, damages produced by robots as a result of errors in manufacturing, assembly, or design usually are producer's liability, but mechanical failures can be caused by not keeping the robot in good conditions, for example, and that is a responsibility of the owner.

In case of user/owner-programmer, the regime is explained previously for damages caused by wrong customization.

About it, the Committee notes that skills resulting from education given to a robot should not be confused with skills depending strictly on its self-learning abilities when seeking to identify the person to whom the robot's harmful behavior is actually due. In this sense, the Committee tries to express its concern about the "bad ideas" that the originator has and that expresses through a robot. The originator is not the programmer or the owner necessarily, but is the person who gives the order to the robot.

That approach of the Committee implies a necessary ethical and moral debate; should the learning capability be limited? Maybe yes, but the "forbidden actions" for a robot should be determined observing each kind of robot. For example, in an ideal scenario where the robot understands the order "do not hurt," a robotic surgeon cannot differentiate between hurting and cutting the skins with a scalpel.

Safety and legality are some of the guiding principles that all producers have to observe, and it is obvious that actions as to steal or to cross the road when traffic is fluid are orders that the robot should never fulfill. So, from a more pragmatic point of view, taking into account the state of the art, it must be determined if the executed order by the robot and the purpose for which is intended are related. A mobile robot with the ability to process orders that implies to make a displacement must be limited to avoid dangerous situations of abuse by the user.

The lack of restraints at the learning ability should be considered as programmer liability. It is without doubt a sign of design deficiency.

5. A potential solution: the Robotic Liability Matrix (RLM)

The model proposed by the Committee observed the learning ability of the robot to distribute the liability between the owner and the manufacturer. In effect, it looks like it is a proposal for robots with the ability to learn only, but the real intent of the Committee is to apply this rule to all autonomous robots. The reason for the confusion is that in this discourse "autonomy" and "ability to learn" concepts are intermingled, forgetting that these two concepts should be analyzed independently. On the one hand, robots can have different levels of autonomy, from a low level for tele-operated robots to a high level for robots with the ability to learn from the environment and react appropriately. On the other hand, also different levels for the ability to learn exist for different robots. For example, a classical industrial manipulator robot has no ability to learn, and an advanced social robot has to learn in order to be able to interact appropriately with humans. So, a robot can be autonomous in a particular task (e.g., a typical industrial task) and has no learning ability (e.g., the classical industrial robots previously mentioned), or has full learning ability and not be autonomous (e.g., a social robot with no ability to move).

Building on this preliminary proposal, our aim is to offer a liability distribution system that allows being applied to any type of autonomous robot, with or without a learning capacity.

5.1. Key factors

In order to ensure reliable results, the next elements have been identified as adequate and robust to endow the Matrix:

Environments. The environment, meaning as the particular situation in which a robot is present, influences on the way in which responsibility is distributed. The more complex it is, the more diligence the producer expects, and the more care the owner must take. In this first stage, we are using the next environment classification system [6]:

- Deterministic or nondeterministic: An environment is deterministic if the next state is perfectly predictable given the knowledge of the previous state and the agent's action.
- Static or dynamic: Static environments do not change, while the agent deliberates.
- Full or partial: A fully observable environment is one in which the agent has access to all information in the environment relevant to its task.
- Single or multiple agent: If there is at least one other agent in the environment, it is a multi-agent environment. Other agents might be apathetic, cooperative, or competitive.
- Known or unknown: An environment is considered to be "known" if the agent understands the laws that govern the environment's behavior.
- Episodic or sequential: Sequential environments require memory of past actions to determine the next action. Episodic environments are a series of one-shot actions, and only the current (or recent) percept is relevant.

- Discrete or continuous: A discrete environment has fixed locations or time intervals. A continuous environment could be measured quantitatively to any level of precision.
- Simulated or non-simulated: In a simulated environment, a separate program is used to simulate an environment, feed percepts to agents, evaluate performance, etc.

Black box equipment. Recording system aboard of the robot is required.

Sensors and actuators. Both are relevant to weigh the risks; the difference between strict liability and negligence of producer depends of the kind of sensors and actuators chosen. Sensors recover data, and actuators act in a consequence; so, the design of the robot must be completely adequate for the tasks. It is not about the quality only but is about safety too.

Mechanical structure. Is the frame of the robot, the skeleton. Safety is not a requirement to individual parts of robots only but is a reference to the whole package too. A dangerous design (cutting edges, heavy materials, or moving components) may affect the risk and harm evaluation.

Learning capability. Referred to real learning capability, and no mere appearance of learning process when the robot is executing a program or accessing to a cloud database where it could download instructions or information. Real learning is the ability to acquire data and elaborate information in order to complete its tasks (Mathias, 2004).

Levels of automation. There are multiple definitions for levels of automation. In order to provide clarity and consistency, we adopt the SAE International definitions for levels of automation, where Level 0 is *no automation* and Level 5 is *full automation* (SAE International).

Depending on the level, the Matrix modulates the responsibility of the producer, the programmer, the owner, or the user using other parameters as the learning capability, the initial knowledge acquired, and the robot architecture.

Human intervention. That factor is really voluble and must be observed carefully. As we said, different kinds of liability can be attributed to different subjects, observing the type of damage and the circumstances of the accident.

In some cases, the final damage can be produced by a concatenation of negligent facts, so it will be necessary to determine how much responsibility each has, and also take into consideration the autonomy of the machine.

5.2. Fundaments

Liability distribution cannot be done observing learning capability and knowledge learned from the owner only, because some software or hardware failures can be produced by the owner.

The Matrix works with identified and isolated accident situations, quantifying the level of implication of each subject in the accident. The accident is not an isolated fact, but is integrated by several and little facts whose consequence is the final accident. Each of one of these facts, which we named *stages*, must bee analyzed and quantified separately. This process of individualization allows us to distinguish between software and hardware failures and human errors.

The stages are checked with an objective questionnaire designed over the user fault principle. If the user fails, the answer to the question is YES. If it is not, the answer is NOT. Each answer has a value of 1, so the final sum of all values determines the degree of involvement of the subject in the accident. Some answers need to be evaluated with reversed values or, as we say, "falsely answered"; it means some YES answers will sum a value of 1 into the NO answer lines and vice versa. That would happen when to keep a *pro consumatore* position could vitiate the results.

Once the analysis of stages is finished, we elaborate the Matrix over the values obtained in each stage, which will show how the liability must be distributed adding up the different values for each subject.

A simplified example of this is presented below.

The RLM has been adapted here to show simple and visually how it works, so that some elements as the types of environments, which have large and complex influence, are not going to be used. The stages have been reduced to one, so the facts happen in a linear manner. There are two tests, one for sensors and other for system. Each test has seven questions and must be remarked that they are oriented to get the liability of the user, so the last questions are about the fault or negligence and extenuating and aggravating circumstances. The questions that must be falsely answered are marked with the letter (F). The producer and the professional programmer on one side, and the owner and the user on the other, will be considered as the same person. The large version of the Matrix would be possible to establish a clear differentiation between each subject, observing the degree and timing of involvement of them.

We are going to use a closed version of Fitorobot. This means that the user cannot customize it. Fitorobot is equipped with different sensors, as a camera for detecting the center of greenhouse corridors using the color as reference, a set of sonars for the same task but when the crop is low, and a touch sensor to stop the activity when Fitorobot hits something. Fitorobot has a bug-testing function that the user must check previously to start the activity. If this test is not realized, Fitorobot does not show any failure signal. The robot has installed a bug-warning system but only works if the bug is produced during the activity. The user has a remote control with an emergency-stop button. The software of Fitorobot is updated to the last version; however, it is supposes that some users have noticed some problems with this update.

5.2.1. *Sensor test*

1. Has there been a sensor failure? (F)
2. If there is a bug-testing function, has it been executed? (F)
3. If there is a bug-warning system, has it been worked?
4. Is there any relation between the presence of the obstacle and the action/omission of the user?

5. Has it noticed fault or negligence?
6. Has it noticed some extenuating circumstance? (F)
7. Has it noticed some aggravating circumstance?

5.2.2. System test

1. Has there been a system failure? (F)
2. If there is a bug-testing function, has it been executed? (F)
3. Is the software updated? (F)
4. Does any error reports about the last update exist? (F)
5. Without the existence of a bug-warning system, would there be fault or negligence in the action-omission of the user?
6. Has it noticed some extenuating circumstance?
7. Has it noticed some aggravating circumstance?

5.2.3. Case 1

The user activates Fitorobot, but he/she does not check the system using the bug-testing function. Fitorobot starts its activity, moving around the corridors. The user has not verified the robot's working environment, so there is a box in the ground and both have the same color. The camera of Fitorobot cannot distinguish between the box and the ground, so it hits the box. The impact blocks the Fitorobot's chain drive causing the robot to cross the corridor, smashing some crops. The bug-warning system works correctly, and the user pushes the remote emergency-stop button after some seconds. **Table 2** shows responses for camera test. The responsibility values for user and producer, computed from the liability index L_j ($j = 1$ for user and $j = 2$ for producer) are given in (Eq. (1)):

$$L_j = \sum_{i=1}^{7} p_i\, V_{ji};\ j = \{1, 2\} \tag{1}$$

where ρ_i is the weight for each question and V_{ji} is given by (Eq. (2))

$$V_{ji} = \left\{ \begin{matrix} 1 & \text{if } Q_i \text{ is marked with X or } Q_i \text{ is marked with F in } (j\, mod 2) + 1 \text{ row} \\ -1 & \text{if } Q_i \text{ is marked with N in row \#1 and 0 in other cases} \end{matrix} \right\} \tag{2}$$

Are $L_1 = 5$ and $L_2 = 1$ in this case.

In this case $p_i = 1$ is chosen for i in {1, 2, 3, 4, 5, 7}, and $p_6 = 0.5$ is used, so that Question 6 does not cancel the full value for a direct question.

	Q1	Q2	Q3	Q4	Q5	Q6	Q7
YES			X	X	X		
NO	F	F				N	X

Table 2. Robotic Liability Matrix (RLM) for camera sensor.

The conclusion is that the damages produced by Fitorobot have been caused by the user's fault largely. The importance of the distribution in this case is that we are not observing the harmful event only, but the whole chain of events from the source to the end. If the user had used the bug-testing function previously, he would have discovered that the sensor was not working correctly. It may seem irrelevant because this does not alter the fact that the box is in the ground and Fitorobot could not have distinguished one thing from the other, but the negligent acts of the user are linked and must be weighed for their computation.

Moreover, it may seem irrelevant (even unfair) that the Question 1 declares the existence of a fail. It attaches one point of guilt to the producer when the accident would have happened in any case. It must be considered that a product is created to work correctly. The fact that the error has no any implication in this occasion does not reduce the responsibility of the producer, who has to guarantee that his/her products do not produce damages.

In the large version of the RLM, it would be possible to determine what extent the answer of Q1 affects to the liability distribution and not attribute a global value only.

5.2.4. Case 2

Taking the case above, we suppose that the touch sensor and the bug-warning system do not work, so the user is not alerted about the crush. Fitorobot keeps moving around the greenhouse smashing the crops. The user is alerted by the noise, and he uses the remote control to stop the robot. The same previous values as in Case 1 for weighing indexes p_i are used in this case.

From RLM for touch sensor (**Table 3**), the liability distribution with respect to this sensor is given by $L_1 = 3$ and $L_2 = 3$.

From RLM for bug-warning system (**Table 4**), the liability distribution with respect to this system is given by $L_1 = 3$ and $L_2 = 3$.

Taking results from RLMs in **Tables 2–4** into account, the user has 11 points of responsibility, and the producer has 7 points. But the most interesting data is the result of bug-warning system test;

	Q1	Q2	Q3	Q4	Q5	Q6	Q7
YES	F			X	X		
NO		F	X			N	X

Table 3. Robotic Liability Matrix (RLM) for touch sensor.

http://dx.doi.org/10.5772/intechopen.69888

	Q1	Q2	Q3	Q4	Q5	Q6	Q7
YES	F		F	X	X		
NO		F				N	X

Table 4. Robotic Liability Matrix (RLM) for bug-warning system.

if the sequence is isolated, it is observed that the producer has more responsibility than the user in the accident. The main reason is that the producer/programmer is responsible about the good work of the internal components. Regarding this issue, the user has to be sure to keep the system updated and to stay informed about the errors published by the producer exclusively.

However, as mentioned above, the Matrix allows observing the damage chain as a whole, so in Case 2, the negligence of the user exceeds in 1 point of the liability of the producer. Into a system of distributive blame, the user is responsible at 61.11% and the producer at 38.89%.

It must be highlighted that the results do not involve that the producer has no any liability because the user is *guiltier*. The Robotic Liability Matrix is designed for distributing the liability, not for attributing it.

6. Conclusions

We have introduced Fitorobot, a successful practical case by the University of Almería, joined to a sample of Robotic Liability Matrix, a new fairer method to distribute the liability to all subjects involved in the use of robots.

In the first instance, the classic theories about product liability are applicable to robots that require continuous human supervision, but in the case of more autonomous robots the standard rules must be improved. The statement approach by the European Parliament, consisting of observing the learning ability to attribute the liability, may not be enough and equitable; all things considered, new products require new solutions.

The result contents in this paper show how the currently product guarantee scheme is obsolete when the product is autonomous in any degree, so the liability for damages does not correspond only to one person.

The Matrix has a competitive advantage over other systems: it will allow defining what would be the content of the guidelines for manufacturers and programmers. A guideline can work as self-regulation system when the Laws are not drafted or passed, and at the same time, it can be used by stakeholders to result in knowledge of the ethical, moral, and legal limits of their work. Guidelines really help reduce time to promote innovation and technology, because they use consensus-based standards, which are recognized by the industry.

The design of the Matrix is under development and improvement process, trying to adequate it to different kinds of robot and situations, but in isolate and laboratory cases, the Matrix is actually working.

Author details

Alejandro Zornoza[1*], José C. Moreno[2], José L. Guzmán[2], Francisco Rodríguez[2] and Julián Sánchez-Hermosilla[2]

*Address all correspondence to: alejandro.zornoza.somolinos@gmail.com

1 Universidad Carlos III de Madrid, Madrid, Spain

2 Universidad de Almería, Almería, Spain

References

[1] Sánchez-Hermosilla J, Rodríguez F, González R, Guzmán JL, Berenguel M. A mechatronic description of an autonomous mobile robot for agricultural tasks in greenhouse. In: Barrer A, editor. Mobile Robots Navigation. InTech; 2010. pp. 583-560. DOI: 10.5772/900

[2] Asimov I. I, Robot. 6th ed. Greenwich, Conn.: Fawcett Publications; 1950. 192 p

[3] Winfield A. alanwinfield.blogspot.com [Internet]. October 30, 2013.Available from: http://alanwinfield.blogspot.com.es/2013/10/ethical-robots-some-technical-and.html [Accessed: March 25, 2017]

[4] Whittaker S. Liability for products. English Law, French Law, and European Harmonisation. 1st ed. Oxford University Press, New York, 2005. 752 p

[5] Avizienis A, Lapire JC, Randell B, Landwehr C. Basic concepts and taxonomy of dependable and secure computing. IEEE Transactions on Dependable and Secure Computing. 2004;**1**(1):11-33. DOI: 10.1109/TDSC.2004.2

[6] Russell SJ, Norvig P. Artificial Intelligence: A Modern Approach. 2nd ed. New Jersey: Prentice Hall; 2010. 1132 p.

Chapter 5

Cybersecurity of Robotics and Autonomous Systems: Privacy and Safety

Francisco J. Rodríguez Lera,
Camino Fernández Llamas,
Ángel Manuel Guerrero and
Vicente Matellán Olivera

Additional information is available at the end of the chapter

http://dx.doi.org/10.5772/intechopen.69796

Abstract

Robots and autonomous systems in general are set to suffer similar cybersecurity problems that computers have been facing for decades. This is not only worrying for critical tasks such as those performed by surgical, or military robots but also for household robots such as vacuum cleaners or for teleconference robots compromise privacy and safety of their owners. What will happen if these robots are hacked? This study presents a survey on the cybersecurity attacks associated with service robots, and as a result, a taxonomy that classifies the risks faced by users when using service robots, distinguishing between security and safety threads, is presented. We also present the robot software development phase as one the most relevant ones for the security of robots.

Keywords: cybersecurity, privacy, safety, robots, autonomous systems

1. Introduction

Practically by definition, all robots are equipped with the ability to sense, process, and record the world around them [1]. In order to offer the best performance, they are continuously gathering information. Under these circumstances, if these robots are compromised, then a two-dimensional security problem arises: first, security issues regarding the virtual side of the robot (data, communications, and so on), and second, those problems associated with physical side that concerns both robot and user integrity. The state of the art present the "Cyber-physical security" term to encompass virtual and physical problems.

Cyber-physical attacks present several challenges that have to be faced [2]. In this study, we have focused in two particular cases: safety and privacy. On the one hand, safety problems cover the consciences associated with the physical integrity of the individuals. People are usually worried about the problems that robots can cause on people or their belongings, hacked robots aggravate these concerns. For instance, there are several commercially available surgical robots as Da Vinci work connected to communication networks allowing remote operations by specialists, what would happen if these robots or their communications would be hijacked? There have been claims of hacked military robots [3], but even service robots at home poses security problems, they could hurt toddlers or produce severe damages to homes (arson, bumping into cars, etc.).

On the other hand, privacy problems associated with robots are spreading in many areas. There are a wide range of service robots that been introduced in homes and retailing spaces. They can be used as mobile teleconference platforms, welcoming assistants, virtual pets, toys, etc. If these robots were hacked, they could provide a lot of private information about the users interacting with the robot or who simply passed by. This information can go from general data (age, size, etc.), private pictures, user routine information, economic, etc., which opens a new stage for cybersecurity of robotic systems [4].

The survey presented in this study overviews the state of the art of the security in mobile robots and it is organized as follows. Section 2 presents a general overview of cybersecurity threats to robots. Section 3 proposes a taxonomy for the risks. Then, in Section 4, the ones related with privacy are analyzed in more detail, whereas Section 5 analyzes the ones associated to safety. Section 6 faces the problems related with software development frameworks for robots. Finally, Section 7 summarizes the work faced in this chapter.

2. Modeling robot cybersecurity threats

Before analyzing in detail the risks generated by compromised robots, it is necessary to model the threads that define the cybersecurity scenario for robots. We propose a model defined overviewing Open Web Application Security Project (OWASP) [5] risk identification and different definitions of cyber-physical security on cyber-physical systems [6]. Our proposal generalizes the previous research, while trying to avoid specifying the taxonomy for any particular scenarios, such as autonomous cars, or social robots.

Figure 1 shows the elements that have to be taken into account to propose a model. We have group the security threads into five groups that will be described later. First, in the top layer, we present the origin of the threats [7, 8]. Threads can have three main origins, which are as follows:

1. Natural, associated to natural disasters.
2. Accidental, generated by the fact that there are no perfect situations as those planned in laboratories.
3. Attack, those events generated by external users with the aim to gain control over a resource of the robot.

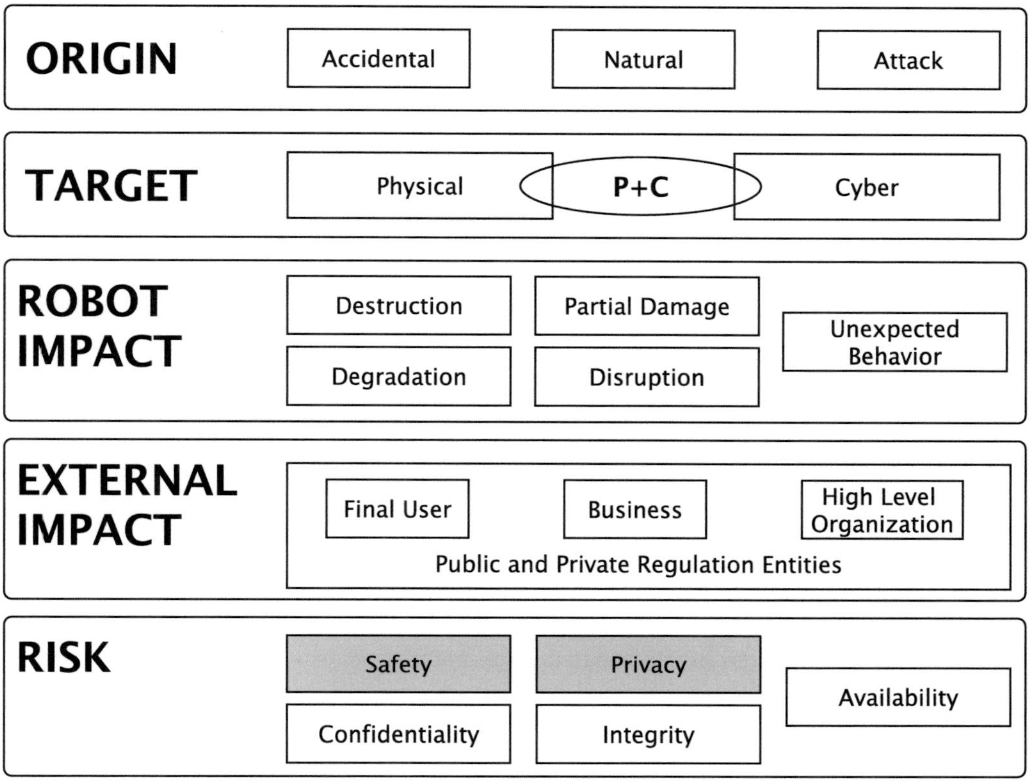

Figure 1. Modeling the robot cybersecurity scenarios.

These elements need to be faced beforehand a robot is deployed in a real environment, and a set of contingency plans should be executed in order to handle these situations.

These plans should consider the three possible targets identified in the second row of **Figure 1**, that is, the robot itself, data managed by the robot, and the combination of both components.

First, threats change the normal operation mode in a physical way. Threads could have been generated by natural conditions, accidental situations, or attacks and could cause five different impacts on the robot (third row in **Figure 1**), which are as follows:

1. Destruction, which implies nonoperability.
2. Partial damage, which involves malfunction of the robot.
3. Disruption, entails the interruption of a single, multiple, and total robot components.
4. Degradation, meaning that the range or capability of any robot component decreases along time.
5. Unexpected behavior, which could be considered as a degradation of the whole robot, not just a component.

Second, cyberthreats can affect the normal operation mode in a virtual way, that is, threads can modify the information gathered, stored, or transmitted by the robot. These threads have more impact on external entities than in the robot itself. In this way, we have organized the impact caused by cyber threats into three groups (corresponding to the fourth row in **Figure 1**), which are as follows:

1. Issues associated to robot manufacturers or open source developers (drivers and core software).
2. Issues associated to third party solutions (libraries) needed by robot manufacturer applications.
3. General vulnerabilities associated with the overall software components of the robot.

Software security issues can be intentionally or unintentionally presented in the software components of the robot. They can be categorized according to [9] into software flaws, security configuration issues, and software feature misuse.

According to existing taxonomies in software vulnerabilities [10], these issues could be added in the analysis, design, implementation, deployment, or maintenance phases. The last one also includes misconfigured robots due to final user adjustments, which allow attackers to get control over the robot.

Third, cyber-physical threats represent the sum up of both previous threads (represented by C+P ellipse in **Figure 1**). From our point of view, it can be illustrated when a sensor or an actuator is compromised. It can be performed by substitution or modification of the hardware, changing them physically or modifying the firmware. The overall system would work in the same manner but the threat would have added a new hidden functionality. The impact associated to an attack to this kind to cyber-physical threat is unexpected; the reason is that the original functional definition has been compromised.

So far, we have defined the robot impact associated with each type of target. Before explaining the risks associated with this model, it is necessary to define the actors that would be involved during or after a security threat happens. By "actors," we mean the set of people and legal entities involved in the deployment of a robotic system.

Our proposal considers four different types of actors: final users, that is, the people interacting with the robot; business users, who deploy a robot for a particular task; robot vendors, which manufacture robots and provide software; and independent software developers who create robot functionalities (individual developers or community developers) for different commercial robots. Additionally, public and private regulatory entities are also involved in the cybersecurity of robotic systems, although they do not use robots directly.

Next section analyzes risks associated with the use of a service robotic platform deployed in private or public scenario (bottom layer in **Figure 1**).

3. Modeling risks

Security risk analysis is usually based on two factors: likelihood of a successful attack against an asset, and the consequence of such an attack [11]. Literature presents several studies about the cybersecurity threats targeted to industrial environments [12]. On the contrary, there are a few studies about the risks associated with service robots at home.

Usual information security overviews classify cyber threats [13] into three fields: confidentiality, integrity, and availability of information. Extended versions for cyber-physical domains [14] add privacy, authentication, authorization, auditability, and nonrepudiation. On top of this, it is also necessary to add safety issues associated with the physical damages caused by a cyberattack.

This section models the risks associated with an attack in terms of the final user type. We identify three groups of final users of service robots such as domestic users, commercial ones, and high-level organization. We set out below the risks that hacked robots pose to these users:

Risks to domestic users can be classified as:

1. Economic risks, which can be quantified as the amount of money required to fix elements of the robot or the environment after an attack.
2. Physical, if any damage to humans happens.
3. Psychological, these risks include the loss of elements people can be attached to, or the dissemination of private information.

Commercial and business risks are related to:

1. Intellectual property, which could be accessed by competitors.
2. Reputation, which could be damaged by the publication of the attack.
3. Economic impact, including the fixing of damaged assets and the loss of profit caused by the damages and their repair.
4. Regulatory problems, if private information about clients or transactions is revealed.

Public Administration (and large very large corporations):

1. Political risks, associated with the loss of confidence by the citizens, not only on the institutions but also in the use of robots in general.
2. Economic damages.
3. National security problems, including the revelation of secrets or damages to strategic assets.

Table 1 summarizes our perception about these threats, which are identify using numbers 1–4 as in the previous description of risks that can be caused by attacks to the physical elements of robots.

The type of asset threatened depends on the different levels of physical attacks to the robot: destruction, partial damage, degradation, disruption, or substitution of their elements, which are presented in the rows of the table.

These assets also depend on the type of user. In **Table 1**, columns correspond to the three different types of users of service robots previously identified (domestic, commercial, or governmental).

Organizations proposed in **Tables 1** and **2** are not oriented to classify the relevance of different risks. They affect distinct actors in different ways, and users' perception about the relevance of these risks is also different. For instance, domestic users of service robotics are usually more worried about privacy issues, whereas corporative users are more concerned about the reputation of the company or its personal reputation in the company.

The same happens with the safety issues; domestic users are more concern about the economic damage that robots can cause in their belongings, whereas corporative users are more worried about reputational damages or potential lawsuits.

From our point of view, attacks on privacy are the one of most relevant risks that service robots could bring both to homes and to business, even more disturbing than the physical

Attack/users	Domestic/personal	Commercial/business	Public administration
Destruction	1, 2	2, 3	2, 3
Partial damage	1, 2, 3	2, 3	2, 3
Degradation	1	3, 4	2, 3
Disruption	1, 2, 3	2	2
Substitution	1, 2, 3	1, 2, 3, 4	1, 2, 3

Table 1. Physical risks for different types of users.

Origin of vulnerabilities/ type of users	Domestic/personal	Commercial/business	Public administration
Manufacturer	1, 3	1, 2, 3, 4	1, 2
Third party	1, 3	2, 3	1
General vulnerability	1, 3	1, 2, 3, 4	1, 2

This table uses the same taxonomy to characterize the risks caused by cyberattacks.

Table 2. Cybersecurity riskS for different types of vulnerabilities.

ones. Service robots are usually not very large robots, so the physical damages they can cause are limited. However, the leak of information can cause severe damages. So, we will analyze focused on privacy characterization in the next section.

4. Privacy risks characterization

Robots can go places humans cannot go, see things humans cannot see, and do things humans cannot do. These characteristics have made them useful tools in many domains as space exploration, rescue missions, hazardous materials manipulation, etc. Their reliability and versatility have also made them very common in industrial environments, such as manufacturing, logistics, etc.

The increase of the computing power and the lowering of the prices of sensors have made possible the extension of robots into other domains. They are entering people's daily life as household robots, autonomous cars, or welcoming assistants in retail business.

It is not hard to imagine why service robots raise privacy concerns [15]. By definition, robots are equipped with the ability to sense, process, and record the world around them. These concerns about privacy problems can go as far as making robotic toys tools for pedophiles: "Because of their built-in cameras, microphones, speakers, and mobile capabilities, all of which can be controlled remotely, mobile Wi-Fi robot toys can pose a risk to families with children when used remotely, an attribute that can be exploited by online pedophiles who want to exploit young children online" [16].

This is not a distant disturbing possibility, it is happening right now. For example, the Wall Street Journal [17] reported that after an investigation into Cayla doll,[1] Germany's Federal Network has issued an order for all parents to find the doll and destroy her. Parents who ignore the order to destroy Cayla could face a fine of up to €25,000 and up to 2 years in prison. On its website, the agency posted a template for a destruction certificate that should be filled in, signed by a waste-management company, and sent to the agency as proof of destruction.

Cayla doll is just an example. Currently, there are many different commercial robots, from vacuum cleaners to social interactive robots that could cause similar problems. From our point of view, the most challenging from the point of view of privacy are the last ones. Social robots usually use cameras to identify users and try to profile them (sex, age, etc.) and also try to figure out their interests for entertaining the users in toys, or for commercial purposes in sales assistants. All this information is private information and any attack on the robot could compromise it.

In the next subsection, we classify the social robots regarding their function, and in the next one regarding the types of sensors they use.

[1]https://www.myfriendcayla.com/

4.1. Types of social robots regarding their use

We distinguish two types of social robots according to the number of potential users interacting with them. First, are the ones designed for personal use. There are two main types of personal robots: nonmobile social assistive robots usually designed as home assistants, as for instance QT[2] or Jibo[3] and mobile telepresence platforms as Beam[4] or Zenbo.[5] The interaction ratio of these robots is 1:*n*, where n is a small number, just the owner, her friends, and family. All of them share some characteristics of their cameras and microphones, the difference between them is the range of the sensors, which is larger in mobile platforms that can move around.

Second group is made up by platforms designed for commercial venues, as for instance, Aldebaran's Pepper. These robots are usually larger than first group ones and also the ratio of interaction in this type of robots is higher (1:*m*), where *m* is larger than *n*, that is, $m >> n$ because these robots interact with the visitors/clients of the venue.

The type of privacy problems is different between the two groups. Robots for home environments, could be listening to conversations or taking pictures of personal info around, or even while you change clothes… Robots deployed in public spaces could learn which are your interests, record conversations between bankers and clients, clone your identification methods, make market studies about your preferences, etc.

Privacy risks have to be classified and should be made explicitly communicated to the buyers of service robots. We propose to classify the privacy risks of service robot according to their sensors. According to this analysis, users should consider what type of robot they are ready to use, and also what kind of security and connectivity they are going to allow in their robots.

4.2. Privacy risks associated with sensors

Roboticists usually distinguish two basic types of robotic sensors: *Exteroceptive* sensors (i.e., lasers, range sensors, cameras, etc.), which provide information about robot workspace, and *Proprioceptive* sensors (i.e., wheel encoders, battery status), which give data about the robot itself. In addition, researches have addressed data fusion problem, that is, how to merge the overall information from robot sensors in just a single flow that could unify all sensors information in one channel.

Table 3 summarizes our proposal of classification for privacy risks, from 1 (very low) to 5 (very high), regarding the severity of the information leakage for the different types of sensors. According to this table, integration of data from various sensors is the most dangerous type of sensor because it can disclose full information about private activities of people, such as their location, images, etc. In an immediate lower level, are people images and conversations, which are the most sensitive leaks in the *exteroceptive* sensors. Finally, *extero-*

[2]http://luxai.eu/products/qt%20robot.html
[3]https://www.jibo.com
[4]https://www.suitabletech.com
[5]http://zenbo.asus.com/

Sensor	*Level*	*Presence*	*Personal information*	*Environment Information*	*Example*
Sensor fusion	5	Yes	Yes	Yes	Full disclosure of scenario activity
Exteroceptive		Yes	Partial	Partial	
Camera	4	Yes	Yes	Yes	Private images of owners
Microphone	3	Yes	Yes	Partial	Recording private conversations or acoustic signals
Range	2	Yes	No	Yes	Recording of private activities
Proprioceptive		No	No	No	
Localization	2	No	Yes	Yes	Maps of the private areas
Encoders	1	No	No	No	Recognize if the robot is moving or not

Table 3. Classification of sensors according to their risks to users' privacy.

ceptive sensors that measure ranges (lasers, ultrasounds, infrared, etc.) are less significant from the point of view of privacy, although some private information can be extracted from range readings. For example, age or sex can be concluded from gait patterns detected using a laser range scanner [18].

In summary, *exteroceptive* sensors are the most sensitive ones from the point of view of privacy if they got hacked. The type of information these sensors were gathering means the relevance of the problem in terms of privacy risks if the robot got compromised.

Information from *proprioceptive* sensors is sometimes considered less sensitive, however it can be used to get information about the current status of robots, which could be used in malicious activities (i.e., planning a housebreaking knowing that guardian robots have to recharge).

Another issue regarding sensors is to clarify where sensor data processing is made. Sensors provide raw data, images for instance. This data has to be transformed into information: detecting a face, identifying that face, or classifying it (age, race, etc.). Algorithms doing these processes are usually very demanding in terms of computer power and manufacturers to safe battery or for lowering prices of robots made this computation "in the cloud." This means that data are sent to remote computers, processed there, and the results sent back to the robot. This may be illegal in some jurisdictions (see previously mentioned Cayla example), but even being legal, these practices pose many risks in the communication (data and information can be intercepted), and also about the storing of this information and its use by the manufacturer or third parties.

Security issues regarding the use of cloud services in cyber-physical system have also been widely discussed [19] in the network community, and their recommendations can be directly applied to robotic systems.

5. Modeling safety issues

We have dealt with the security problems; cyberattacks to robots also raise safety concerns. We propose to classify the safety problems associated with robots attending to the behavior of robots observed by their users. We distinguish two basic behaviors: controlled or noncontrolled, that is, if the robot is having a responsive behavior or if it is acting randomly. This is an external and subjective perception, but we think that can be easily asset by any robot user.

In **Table 4**, we distinguish between two types of anomalous situations that a service robot can suffer and regular behavior of the robot. We propose that the situation has to be appraised by external observers and classified in one of these three categories: normal, abnormal, or under attack.

"Normal" behavior is the expected behavior of the robot according to the manufacturer specifications and user commands. By "Abnormal," we mean that the robot is not working as expected but it is not under attack. If the observed behavior of the robot is controlled, that is, responding to stimuli, the main problem is that the robot will not be able to fulfill its task. If

Situation	Behavior	Status	Threads
Normal	Controlled	Regular	
Abnormal	Controlled	Tasks are not performed in the designed way, but the robot reacts to stimuli	Lack of completion of actions commanded by legitimate users of the robot
	Noncontrolled	Tasks are not performed in the designed way and the robot behaves randomly	Potential damages to users/ environment. Lack of completion of actions commanded by legitimate users of the robot
Under Attack	Controlled	A new task is performed The same task is performed	The robot could be used as a weapon Lack of completion of actions commanded by legitimate users of the robot
	Noncontrolled	Tasks are not performed in the designed way and the robot behaves randomly	Potential damages to users/ environment. Lack of completion of actions commanded by legitimate users of the robot

Table 4. Classification of safety problems in robots associated to their behavior.

the robot is not responding, more severe damages can happen. In both cases, one usual solution is to reset robot using the mandatory reset physical button [20] available in every service robot.

The "under attack" case is the most difficult to assess by a user. It is difficult for an external observer to know what is happening inside the robot. Observers can verify whether the robot is "controlled," that is, it is responding to stimuli or not. If it is, but is doing a different task than expected, this could be an indication of an attack. But even if the robot is working as expected, it could have been compromised and the attacker could be hiding under the normal behavior of the robot.

Regarding the severity of the safety risks, the main features of robots that define the consequences of safety problems are based on physical dimensions of the robot, mainly size and weight of the robot; and also, the hazardous elements accessible by the robot, which can be components of the robot (batteries, effectors, etc.) or elements that can be reached by the robot (cargo being delivered, environment elements being manipulated, etc.).

Thus, we propose a classification of safety risks for robots based on the size of the robot. We propose a three-level classification of risks, which are as follows:

1. Low-level risks: Robots that can be carried out by users. This means that robots interacting with children or handicapped people should be small enough to be considered at low risk.
2. Medium-level risks: Robots that cannot be carried by users, but whose size is smaller than their users. This means that the robot cannot be handled by users, but they are not intimidated by them.
3. High-level risks: Robots which are larger than the users in any of the user dimensions like size, height, and speed.

5.1. Examples of attacks

There are different ways of attacking a robot, and different levels in the severity of the intrusions. **Table 5** shows some examples of cyberattacks to service robots indicating if it generates a privacy [1] or a safety problem [2].

"Stealth attacks" can be implemented in different ways. In this type of attack the attackers basically try to modify the sensors readings of the robot to induce an error. This can be achieved by modifying the environment, or interfering with the sensors. Some solutions could have been proposed to detect this kind of attacks for instance using the cumulative sum (CUSUM) [21] to detect errors in a range sensor readings, which could cause collisions in a mobile robot.

Another well-known attack is the "replay attack." If attackers are able to intercept the communications of the system, they can replay captured packages, even if they are encrypted. If the communication protocol is not prepared for this kind of attack, the system will consider these replayed packages as legitimate, and make mistakes in the decisions. This type of attack is also used as a precursor to discover the interiority of the system, looking for new weaknesses.

Attack	*Type*	*Example*
Stealth attack	2	Modification/Substitution of sensors readings
Replay attack	1,2	Attacker impersonating roles
Covert attack	1	Third party applications sharing personal data
False-Data injection	1,2	Medical robots
DoS attack	2	Robot not working at all
Remote access	1,2	Robot controlled by an attacker
Eavesdropping	1	Attacker monitoring robot-user messages

Table 5. Example of attacks to service robots.

There are other types of attacks to robots not related to their sensors. They can be targeted to the cognitive elements of the control system. For example, in a medical robot, if false information is given to the system about the condition of the patient, the robot could take wrong decisions that potentially could cause severe damages. This kind of "False-Data Injection" could also be used in other type of robots, for instance, providing false maps to a mobile robot to bring a collision, or fake information to a robotic shop assistant to mislead clients.

From the point of view of privacy, "Eavesdropping" is one of the most feared threads in computer systems. The same concern applies to robotic systems. If robotic systems exchange information with other off-board systems, this communication can be compromised and private information about the users can be obtained (this problem has been described in Section 4.2).

Denial of Service (DoS) is other classic type of attack. DoS attacks in robotics generally mean that the robot stops working, so damages are not suffered by robots, neither robots damage people or their environment. Damages are due to lack of the service provided by the robot. The severity of the attack depends on the criticality of the service to be supplied.

A worse case arises when the robot is not just stopped but hijacked. This is known as in cybersecurity as a "Remote Access." In this situation, robots pose safety problems, not just privacy ones.

Classifying the severity of the attacks is a challenging problem. It is very difficult to predict the consequences of the attacks. Even a small loss of data can have catastrophic effects on the reputation of a company. A single private image eavesdropped from a home robot can be used for blackmailing a tycoon. DoS attacks that only prevent the robot for doing its work could look less dangerous, but they are really important, for instance, in tele-surgical robotic systems [22].

6. Securing the development

As we have previously mentioned, most of the cybersecurity problems in robotics are due to the lack of awareness among developers of software for robots [4]. Software controlling robots need to be secured, which means that the methodologies, tools, and development frameworks used have to be secured.

Different robotic development frameworks (RDF) have appeared in last years. RDFs simplify and speed up the development of robotic applications because they ease the portability of applications among different robots, favoring code reusability and reducing the cost of new developments.

Major contributions of RDFs are hardware abstraction, which hides the complexity of managing heterogeneous hardware by using standard interfaces; distribution of computing resources, which let programmers spread the computation of complex systems over a network; and the creation of "communities" of developers around them for sharing code, tools, etc.

However, contributions provided by RDFs have also brought security flaws to the robotics ecosystem. For instance, sharing code without the appropriate precautions open a door for *malware* infections. In the same way, using distributed computation means that data is being transferred among different computers, in many cases using public infrastructure (public Wi-Fi networks) where different types of attacks can happen, as we have previously described.

This is due to the youthfulness of the robotic community. In many cases, transference from research and university environments to commercial products is happening too fast. Mainstream RDFs do not have any security mechanism in their design.

For instance, robotic operating system (ROS), the "de facto" standard for robotic systems, was designed without taking into account almost any security protections. Malicious malware can easily interfere with ROS communications [23], read private messages or even supersede nodes.

Cybersecurity was not a requirement for ROS at its conception because it was designed mainly for research purposes, but now it has to be because it is being used in commercial products. The new version of this framework (known as ROS 2.0) which is currently in development (beta stage) is trying to solve these problems by using DDS as a standard middleware [24].

RDFs are based on software engineering principles, which means that they are suitable for adapting secure methodologies [25] used in more mature software development communities.

Using standard methodologies and secure tools is not a new issue, it has been proposed several times [26] but as robots use is expanding to homes and streets, with people increasingly trusting on them for more tasks of their daily life, this need is becoming imperative.

7. Conclusion

In this chapter, we have dealt with the security and safety problems potentially caused by cyberattacks to robots. As major contribution, we have proposed a taxonomy for the attacks to service robots. This classification takes into account different types of assets that can be compromised, which could be physical or immaterial (information, reputation, etc.) and could be useful when deciding whether to acquire a robot or not.

We have classified the different types of sensors that usually equip service robots and the relevance of the information they can gather in case of a cyberattack. This information could also be useful when configuring robots for different uses.

We have also analyzed the safety problems that arise when a robot is compromised. In this case, we propose a taxonomy of the problems based on the subjective perception of users about the behavior of robots and a classification of the risks depending on the size of the robots.

Finally, we have analyzed the role of the development frameworks used to program robots. In this case, we recommend the use of standard methodologies and good practices common in other software development environments. We also point out that some well-known middleware, as ROS, have severe security flaws that should be taken care of if it is going to be used for commercial products.

Acknowledgements

This work has been partially funded by INCIBE (Instituto Nacional de Ciberseguridad S.A.), a public company of the Kingdom of Spain and the Universidad de León under the Adenda 21 of the framework agreement between the two entities.

Author details

Francisco J. Rodríguez Lera[1], Camino Fernández Llamas[2], Ángel Manuel Guerrero[2] and Vicente Matellán Olivera[2]*

*Address all correspondence to: vicente.matellan@unileon.es

1 AI - RoboLab University of Luxembourg, Belval, Luxembourg

2 RIASC - Research Institute on Applied Sciences to Cybersecurity, Universidad de León, León, Spain

References

[1] Calo RM. Robots and privacy. In: Bekey P, Abney G, Lin K, editor. Robot Ethics: The Ethical and Social Implications of Robotics. Boston: MIT Press; 2011

[2] Humayed A, Lin J, Li F, Luo B. Cyber-Physical Systems Security - A Survey [Internet]. 2017. arXiv preprint. arXiv:1701.04525

[3] Rawnsley A. Iran's Alleged Drone Hack: Tough, but Possible [Internet]. 2011, Wired. Available from: https://www.wired.com/2011/12/iran-drone-hack-gps [Accessed: March 1, 2017]

[4] Apa L, Cerrudo C. Hacking Robots Before Skynet. Seattle: IOActive Inc.; 2017. pp. 1-17. https://ioactive.com/pdfs/Hacking-Robots-Before-Skynet.pdf

[5] v.40, OWASP Testing Guide. OWASP Testing Guide.OWASP. [Online] [Internet]. 2014. Available fromhttps://www.owasp.org/images/1/19/OTGv4.pdf [Accessed: April 1, 2017]

[6] Alvaro A. Cardenas; Saurabh Amin; and Shankar Sastry. Secure control: Towards survivable cyber-physical systems. In: IEEE Distributed Computing Systems Workshop, 2008 (ICDCS'08); 2008. IEEE; 2008. pp. 495-500

[7] Gabriel J. Mission-centricity in cyber security: Architecting cyber attack resilient missions. In: 5th International Conference on Cyber Conflict (CyCon); 2013. pp. 1-18

[8] Conklin A, White GB. E-government and cyber security: The role of cyber security exercises. s.l:. 39th Annual Hawaii International Conference on System Sciences (HICSS'06); 2006. IEEE; 2006. pp. 76b–79b.

[9] LeMay E, Scarfone K, Mell P. The Common Misuse Scoring System (CMSS): Metrics for Software Feature Misuse Vulnerabilities. National Institute of Standards and Technology (NIST); 2012. Interagency Report

[10] Piessens F. A taxonomy of causes of software vulnerabilities in internet software. Proceedings of the 13th International Symposium on Software Reliability Engineering; Los Alamitos, CA: IEEE Computer Society Press; 2002

[11] Byres E, Lowe J. The myths and facts behind cyber security risks for industrial control systems. Proceedings of the VDE Kongress. 2004;**116**:213-218

[12] Ten CW, Manimaran G, Liu CC. Cybersecurity for critical infrastructures: Attack and defense modeling.. IEEE Transactions on Systems, Man, and Cybernetics-Part A: Systems and Humans. 2010;**40**(4):853-865

[13] Cavelty MD. From cyber-bombs to political fallout: Threat representations with an impact in the cyber-security discourse. International Studies Review. 2013;**15**(1):105-122. DOI: 10.1111/misr.12023

[14] Yan Y, et al. A survey on cyber security for smart grid communications, s.l: IEEE. IEEE Communications Surveys & Tutorials. 2012;**14**(4):998-1010. DOI: 10.1109/SURV.2012.010912.00035

[15] Denning T, Matuszek C, Koscher K, Smith JR, Kohno T. A spotlight on security and privacy risks with future household robots: Attacks and lessons. s.l. In: 11th International Conference on Ubiquitous Computing (UbiComp '09); 2009. ACM; 2009. pp. 105-114

[16] Yong S, Lindskog D, Ruhl R, Zavarsky P. Risk mitigation strategies for mobile Wi-Fi robot toys from online pedophiles. s.l. IEEE International Conference on Privacy, Security, Risk and Trust 2011 and IEEE International Conference on Social Computing 2011. IEEE; 2011. pp. 1220-1223. DOI: 10.1109/PASSAT/SocialCom.2011.194

[17] Andrea Thomas, Germany issues kill order for a domestic spy - Cayla the toly doll. Wall Street. Wall Street Journal. [Online] [Internet]. 2017. Available from: https://www.wsj.com/articles/germany-issues-kill-order-for-a-domestic-spycayla-the-toy-doll-1492098755 [Accessed: April 28, 2017]

[18] Palleјà T, Teixidó M, Tresanchez M, Palacín J. Measuring gait using a ground laser range sensor. s.l: MDPI. Sensors. 2009;**9**(11):9133-9146. 10.3390/s91109133

[19] Yana Z, Zhangc P, Vasilakosd AV. A survey on trust management for Internet of Things. s.l.: Elsevier. Journal of Network and Computer Applications. 2014;**42**:120-134. DOI: 10.1016/j.jnca.2014.01.014

[20] ISO/TC 299 Robotics. Robots and Robotic Devices -- Safety Requirements for Personal Care Robots. Geneve: ISO, 2014. p. 79. Standard

[21] Sabaliauskaite G, Ng GS, Ruths J, Mathur A. Experimental evaluation of stealthy attack detection in a robot. s.l. In: IEEE 21st Pacific Rim International Symposium on Dependable Computing (PRDC). IEEE; 2015. pp. 70-79. DOI: 10.1109/PRDC.2015.33

[22] Coble K, Wang W, Chu B, Li Z. Secure software attestation for military telesurgical robot systems.. MIlitary Communications Conference (MILCOM 2010). San José, California: IEEE; 2010. DOI: 10.1109/MILCOM.2010.5679580

[23] Lera FJR, Balsa J, Casado F, Fernández C, Rico FM, Matellán V. Cybersecurity in Autonomous Systems: Evaluating the performance of hardening ROS: s.n. XVII Workshop en Agentes Físicos; 2016; Málaga (Spain). 2016. pp. 47-54

[24] Woodall, William. ROS on DDS. *ros.org*. [Online] [Internet]. 2016. Available from: http://design.ros2.org/articles/ros_on_dds.html [Accessed: Mayo 10, 2017.]

[25] Finnicum M, King ST. Building secure robot applications. In: Proceedings of the 6th USENIX Conference on Hot Topics in Security, HotSec'11. Berkeley, CA: USENIX Association; 2011. pp. 1-1. http://dl.acm.org/citation.cfm?id=2028041

[26] Tomatis N, et al. Designing a secure and robust mobile interacting robot for the long term. IEEE International Conference on Robotics and Automation (ICRA); Taipei, Taiwan. 2003. DOI: 10.1109/ROBOT.2003.1242256

Section 4

Ethical Impacts

Chapter 6

Ethic Reflections about Service Robotics, from Human Protection to Enhancement: Case Study on Cultural Heritage

Luca Giuliano, Maria Luce Lupetti, Sara Khan and Claudio Germak

Additional information is available at the end of the chapter

http://dx.doi.org/10.5772/intechopen.69768

Abstract

In a vision of future implications of human-robot interactions, it is vital to investigate how computer ethics and specifically roboethics could help to enhance human's life. In this chapter, the role of design expertise will be emphasized by setting multiple disciplines into a constructive dialogue. The reflections will take into consideration different themes, such as acceptability and aesthetics, but above all the ability to generate value and meaning in different contexts. These contexts could find a description in the concept of human enhancement, connected through each other with the skills of the design research. The methodology of the design research will find applicability in the case study of Virgil, where a roboethic approach is contextualized into a cultural heritage field. In this field, it is shown how the ethical approach will bring a benefit to local communities, but at large to any social and cultural strategies involved in the stakeholders' network.

Keywords: roboethics, service robotics, design research, cultural heritage, human-robot interaction

1. Introduction

This chapter presents a reflection about the future implications of human-robot interaction. It is a common belief, in fact, that robotics, especially the service category, will have a great impact on many aspects of our society in the near future. Society is actually experiencing what Łichocki et al. [1] defined as a "robotic demographic explosion." Given the complexity of this emerging society, the interdisciplinary cooperation between different disciplines will play a strategic role in this coming era of human-robot coexistence [2].

The development of service robotics implies that, very soon, human and robots will need to learn how to share not only environments but also activities. This might result in competition between the two subjects, as in the case of industry automation and the issue of job loss [3]. In this scenario, roboticists will play a decisive role in the establishment of the boundaries of this competition. Being aware of this responsibility will also shape the future directions of this discipline.

The theme of human replacement in many activities is already widely addressed, both from scientific community [1, 2, 4] and from mass media [5–7]. Furthermore, the relationship between man and machine has always fascinated science fiction writers who often enjoyed imagining futuristic scenarios where robots affirm their domain. These dystopian futures, over the time, favored a general negative attitude of people toward robots, who were mostly described suspiciously by the writers.

The scientific debate, instead, has always questioned which effect might bring this scientific and technological progress on productive systems and, in particular, on human activities. The main position statements, from the scientific/humanistic perspective, are represented by two conflicting ways of thinking. On one hand, the technological progress of robotics is considered a positive phenomenon, whereas on the other hand, there is skepticism against the diffusion of robots.

Regarding this skepticism and the main fears, an issue is represented by the fact of not knowing where the technological progress is leading: the concept of technological singularity [8], namely, the civilization status where the technological progress accelerates beyond the human capacity of understanding and forecasting its behavior. However, some experts stated that there is no need to worry, since the diffusion of new technologies has always created new jobs for people, while replacing the old ones [9]. The result, then, is a displacement rather than a replacement.

Therefore, a possible answer could be to rethink the human role, by redesigning its behaviors and life style. It is necessary, then, to investigate the key aspects of these new conditions. If machines can perform a work better than human, it is not reasonable to not allow it to do that work. Humans, then, should take this as an opportunity to do/become something else. Regarding this, Brynjiolfsson and McAfee affirmed that "there's never been a worse time to be a worker with only 'ordinary' skills and abilities to offer, because computers, robots, and other digital technologies are acquiring these skills and abilities at an extraordinary rate" [10].

Thus, the theme of competition in work environment is still a hot topic in the debate about human-robot interaction. This is due to the fact that, even if machine were developed by humans to support them in risky, harmful, and repetitive works, the development of artificial intelligence is nowadays raising the risk of excluding people from productive processes. However, many researchers believe that this statement is not valid anymore. Frey and Osborne [11], in fact, highlighted three main kind of human tasks for which there is no risk of replacement from computing and robotics, at least in the short term. These three categories consist of tasks that require a high sensorial perception and manual dexterity, creative tasks, and tasks that require social intelligence. According to the authors, these tasks are still not replicable by artificial intelligence, since it is not possible to design a software that is able to equate sensorial perception or manual dexterity, and because the psychological processes that characterize creativity and social intelligence are difficult to be specified. As a matter of fact, even machine learning algorithms based on big data are unable to codify certain human processes, such as negotiation, persuasion, or concern, that are required in certain tasks. These assumptions drove Frey and Osborne to redefine trades and professions on the basis of the previous tasks categories. The more these tasks are relevant for a work, the more the related profession needs to be prevented from the risk of automation.

All these reflections highlight that technological development should not be the sole driver for the future of society, but rather, new common strategies and methodologies are required to manage the relationship between humans and machines. One of these development strategies is represented by *roboethics*. This term, coined by Veruggio [12], stands for a discipline that aims to establish the basis of human-robot relationship. This discipline is based on two main principles: the dialogue between all the actors involved in a project, and the creation of relationships among these actors for achieving shared solutions. From the service robotics, and the human activity replacement, point of view, roboethics is important, because it does not provide answers, but it rather generates the conditions that, in a project, can foster the appropriate questions about the impact of a robot both on society and the territory.

The academic debate that is developing around the use of robotics in everyday life has therefore attempted to propose a reflection about ethics and work of the future. The discussion is designed on competitive factors between human beings and machines such as cross-cultural competencies, transdisciplinarity, and adaptive thinking [13]. In general, at this precise moment in the history, the paradigms of machine usage are changing, as well as the motivations of their production. There is a shift from machines built to protect and preserve humans, to machines that can enhance human abilities, giving to the human itself a new meaning.

2. Human enhancement

Gerd Leonhard [14] affirmed that it is correct to exploit technology, as long as people are not addicted to technology themselves. In fact, on one side, digital systems are getting more and more efficient, whereas on the other side, people are losing some human characteristics. Regarding this, the futurist Leonard coined a neologism: androrhythms. This term is meant for describing what is particularly relevant for people: human's rhythms, not machine's rhythms, namely algorithms. The androrhythms includes human aspects, such as empathy, compassion, creativity, and storytelling. The risk, unfortunately, is to lose these rhythms in favor of automation [14].

The main impact that robotic technologies might have on humanity can be associated at three main levels: activities, environment, and relations. Roboethics should take into account all these levels not only by addressing all the possible negative consequences but also all the opportunities to enhance humanity and create value [15]. Regarding human activities, for instance, robotics could be used to support existing tasks by providing new tools or by replacing existing tasks for people while providing them new tasks. From the environmental point of view, robotics may be used to replace people in unsafe environments, to prevent environmental damage from human, or to provide more effective tools for environmental care, such as restoration, or energy management. Finally, concerning relations, robotics might be an opportunity for connecting people through remote embodied interaction, or it could be used to promote social behaviors in people with special needs, such as autistic children, hospitalized patients, or elderly [16].

In the book "The second machine age," Erik Brynjolfsson and Andrew McAfee try to predict the future relations between human jobs and robotics [10]. Many tasks that human find easy and natural to execute in the physical world are tough to be managed by a robot. As per Brynjolfsson and McAfee, the working classes that are going to be replaced ruthlessly by this technology evolution are analysts and market expert, because their repetitive working methodology of analysis is going to be substituted by an algorithm managed by an artificial intelligence. Concerning physical activities, humans have more flexibility in respect to machines. Automate a single work task activity, like solder a wire or put a screw, it is relatively easy except if the machine operates in a controlled environment and all the passages that the robot has to progress are clear and that is why in the production chain, machines are always overseen by humans. Neumeier [17] defines this as a robot curve of a job. "The Robot Curve shows that pushing capabilities down the curve produce profits. Every time a new idea become a professional practice or a professional practice becomes a rote procedure, or a rote procedure becomes a robotic operation, there's a chance for someone to profit" [17]. Neumeier did not see the robotic revolution as a closure; for him, the robot curve is a waterfall of opportunity that flows from the creativity to the automated. For the researcher, the humanity is in a recession, because we are confusing cause and effect and we are trying to apply industrial age ideas to robotic age realities, and the result has been a creative and economic vortex. As designers, we have to start to rethink the future work and give to the future worker new skills, not in competition with the ones of the machines. In this sense, the Institute for the Future IFTF, an independent, nonprofit research organization, in 2011, published the report Future Work Skills 2020 [18] (**Figure 1**).

In the report, both the characteristics of the future work and the main technology driver to achieve those results are described. In conclusion, the demographic grown run faster than the grown of the workplace and the productivity is no more related to the occupation. Martin Ford, the author of the book "rise of the robot," underlines how the automation may lead to a global unemployment: millions of workers will be out from the labor force without the possibility of getting back into it. The question is: What will these people do? There is no an absolute answer to this issue, but today, humanity can count on an incredible set of knowledge. We have a huge baggage of data and information, crossing them, we could have the possibility to solve problems that we drag from centuries [19].

http://dx.doi.org/10.5772/intechopen.69768

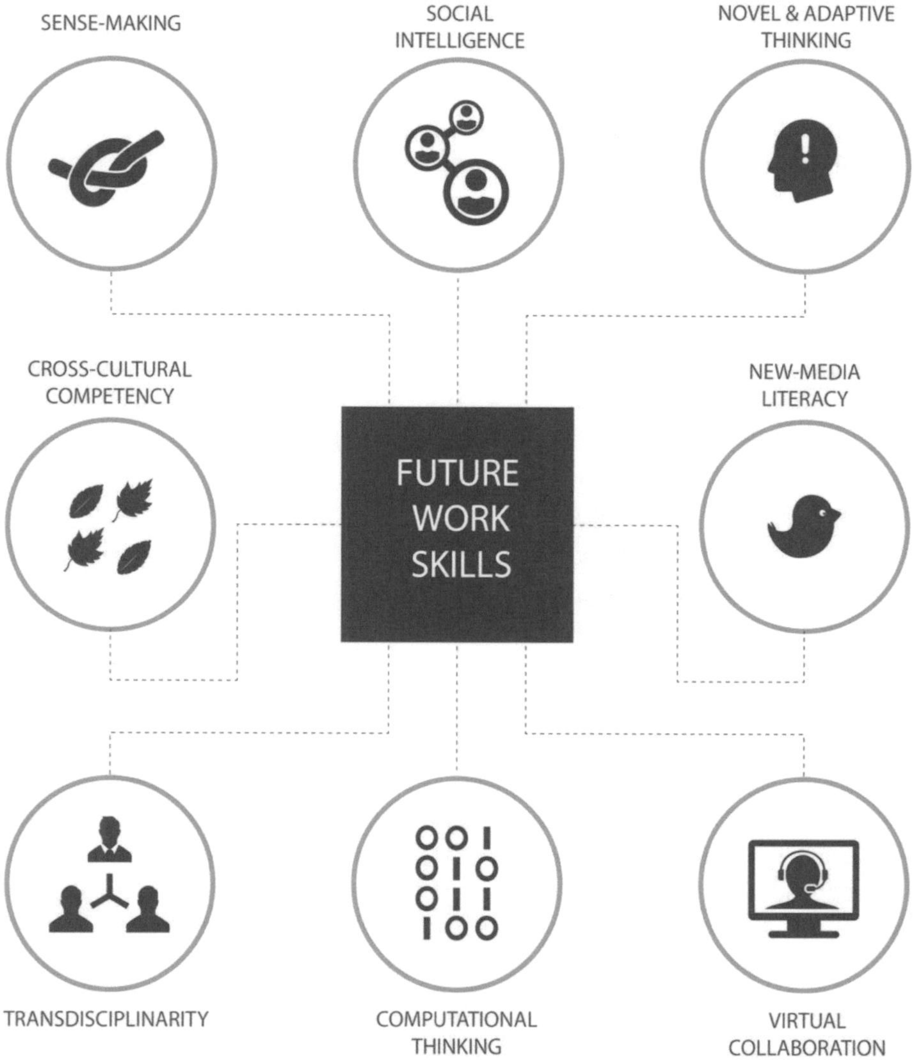

Figure 1. Future work skills referred to the work by Davies et al. [18].

3. The role of design research

Recently, different competences were conveyed in the robotic field. If robotics was firstly a primary field for mechanical and electronic engineering, nowadays multiple discipline and knowledge contribute to it. The evolution of robotic artifacts led researchers to question themselves about different themes, such as acceptability and ethics. For this reason, it was necessary to extrapolate the debate to different disciplines: design is just one example. Through the contemporary design discipline, in fact, roboticists extend their ability to generate value and meaning, creating relationships among technology, human needs, and contexts. Accordingly, the design discipline can contribute to the robotic field at three main levels: artifact, stakeholders, and context.

The design actions that focus on the *artifact* level consists mainly in the application of design practice methods for the development of novel robots. In this regard, Luria et al. [20] provide a further description of the aspects addressed through the design process, especially in the case of social robots. They explain that designers are asked to face three interrelated robot's aspects, namely, morphology, nonverbal behaviors, and interaction schemas [20]. Many case studies present detailed descriptions of robot's design processes, in which it is possible to identify some common actions. The first design ideas are usually explored and shared through sketches and 3D models, as shown by Refs. [20–22]. Then, a key role is played by the prototyping actions. From low-fidelity to high-fidelity prototypes, tangible artifacts, which appear or behave as desired, allow fast testing and iterations, as reported by Šabanović et al. [23]. Finally, usability testing represents another common action in the design of novel robots. Vandevelde et al. [24], for instance, developed a robotic toolkit, that is easy to build, by doing multiple design iterations and regularly testing with non-expert users.

Given the focus on practice-based methods for the design of artifacts, it is possible to state that at the first level, the design action is characterized by the adoption of the Research through Design (RtD) approach [25]. In fact, even though none of these examples make explicit reference to RtD, they all employ "methods, practices, and processes of the design practice with the intent of generating new knowledge" [26] which perfectly falls in the definition of RtD given by Zimmerman and Forlizzi.

RtD, however, is not limited to the employment of design practice methods. As also mentioned by Zimmerman and Forlizzi, RtD emerged from different design approaches, such as participatory design and critical design, that go beyond the artifact in favor of a deeper understanding of human perception, emotional reactions, and emerging behaviors. These aspects introduce the second level to which design research can contribute: stakeholders.

As *stakeholders* are intended *any entity who can affect or is affected by a project, referring to the definition by* Freeman [27]. As a matter of fact, the efficacy of each project results on one hand from motivations, visions, and methods of who develop a project, whereas on the other hand, it depends on many user acceptance factors. Given this fact, it becomes crucial to identify all the actors who can potentially interact with the project. A key action for that is represented by stakeholders mapping [28], which results in graphical visualizations that allow to increase awareness on implications and consequences that a project might have [29].

Other key design actions, focused on the stakeholders, address more specific aspects of perception and people's attitude toward robots. In particular, participatory design methods, such as interviews, questionnaires, hands-on workshops, etc., are often adopted with various aims. They can be employed to get knowledge about different aspects of robot's acceptability, such as in the questionnaire-based study by Choi et al. [30] aimed to identify positive and negative aspects of edutainment robots according to parents. In other cases, these can be aimed at observing emerging interactions, for example, an ethnographic study with users performed in elderly care center in Japan [31]. Furthermore, participatory design can engage creatively the stakeholders: to cocreate robotic solution that in this way results from the mutual shaping between society and technology [32].

Participatory design actions can also be focused on the understanding of physical and socio-cultural factors that determine the specific nature of a *context*. This third level of design actions, in fact, is often characterized by actions like context mapping [33], immersive investigations, interviews, and direct observations, which aim at investigating the current scenario, as well as to develop design proposals, that usually take the form of design scenarios [34] or storyboards.

It is evident that these three main levels of design actions are deeply interconnected. Every design project, indeed, affects all the factors mentioned above. Even though some projects might focus on more specific aspects of human perception, whereas others more on contextual challenges, every project has some implications at all the three levels. Every design project, in fact, results from the simultaneous investigation of four key assets: form, function, value, and meaning. Speculations about the form, which can be considered the traditional matter of design, takes shape from the combination of creativity with technical feasibility and new technological opportunities. Function, instead, results from the meeting of technology with actual needs. The value is created at the crossroads of economic profit and the humanities' search for interest. Finally, meaning arises out of the encounter between the hermeneutics of humanist culture and the intuitions of art. These aspects, specific of the design culture, match with roboethics, and the design methodologies represent a valuable tool to develop acceptable robotic solutions.

The project Virgil, a telepresence robot for cultural heritage, was developed by taking these considerations on ethics and the role of design research. The project was aimed at achieving an ethical solution by addressing the artifacts, all the possible stakeholders, and the specificity of the contexts.

4. Virgil, a case study with a roboethic approach

Virgil [35] is a project that was conducted by Politecnico di Torino in collaboration with the TIM Jol CRAB, a private research lab focused on cloud-robotics-related projects, among which is the telepresence robotics. The research, developed with applicative purposes, represented a good chance to apply the methodologies of roboethics design. Starting from an in wild experience set up in the Racconigi's castle, a territorial museum inserted in the cultural heritage of the Piedmont; the primary goal of the project was to develop a robotic service application for cultural heritage (**Figure 2**).

This experience, drawn up by participatory design approaches, was aimed to enrich the museum visiting experience, through a digital tool that increases the interactivity of the visit. At the beginning of the design process, shared ethical reflections were made through the iterative

Figure 2. Racconigi's Castle, Italy.

dialogue between the design team and the stakeholders. Four important ethical guidelines have been highlighted. Firstly, the robot has to enhance the work of the museum guide and does not be competitive with it (technology as support). Second, the telepresence operability of the robot makes it capable of being moved and show in real time the inaccessible area for people of the museum. Third, the robotic solution does not have to overstructure the environment and spoil the artistic aura of the cultural heritage. Fourth, the robot can overtake the issues related to the architectural barrier and make the whole area accessible for people with motion disability (**Figure 3**).

According to the roboethics reflections, the main guideline followed during the project was to avoid the human work replacement of the museum guide. Furthermore, instead of replacing the human work, it was enhanced by providing a novel tool, together with new skills and new interaction opportunities. Applications of robots for museum purposes can be resumed in three categories: as a museum guide, in telepresence, like installation. For which concern robot used like a museum guide, the two model samples taken in consideration in the developing of the project were TPR-Robina and Robot Norio. TPR-Robina is a typical example of use of robot in substitution of the human work. The robot was used in Kaikan's museum [36] in early 2000 for welcoming and routing the visitor. The application was not successful because of issues related to the interaction between the people and the robot, it was basically too slow when answering the visitor's questions. Robot Norio, on the contrary, was used to enhance the capability of a disabled guide to conduct in a remote way a tour of the visit at Château d'Oiron [37].

The solution presented with Virgil's project stays in between the two samples. It is possible to say that nowadays, this kind of application presents some deficiencies regarding adaptability to the tour of visit, aesthetic coherence, and easiness of communication with the visitors (in particular robot as museum guide).

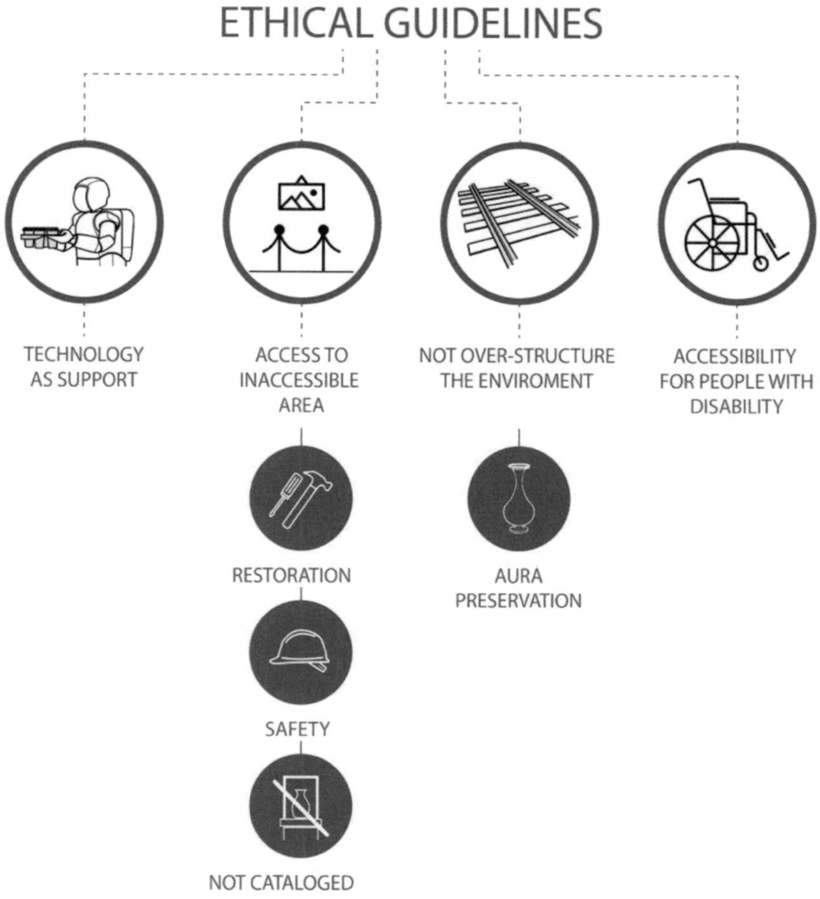

Figure 3. Ethical design guidelines for cultural heritage.

From these considerations, the robot Virgil, a tool for museum's guides to enhance their cultural storytelling, has been developed.

Between the stakeholders involved in the project, museum guide is one of the figures that will benefit from the project. The museum guide, in fact, embodies the ambassador of the knowledge and it is a vassal of the museum identity. The guided tour is the learning tool par excellence, one of the most traditional ways for educating the visitors and one of the pedagogical instrument more direct and efficient used for developing the understanding and comprehending the artifacts exposed. A universal pattern of exposition of the guided tours does not exist; every museum presents a different scenario of exposition: different collection, different spaces, a different narrative ability of the guide, and different composition of the group of visiting.

In the project development, the work of the museum guide it has been analyzed very carefully to understand desiderata and planning of the tasks. Traditionally the museum experience is an active communicational process between the museum (broadcaster) and the visitor (receiver) through an artifact (medium) [38]. The work of the museum guide is facilitating this relation, accompanying and stimulating the communicational process. The pillar on which is based the work of the museum guide is the dialogue within the visitors; an inspiring and interactive exchange composed of a succession of questions and answers could create a very useful tool for the learning and could allow to visitors to give a personal interpretation about what they are watching.

Visitors did not go to the museum as an "empty container," they bring with them their beliefs, their knowledge, and their culture, so they have not to be handled as a passive receiver of the information but as actors dynamically enrolled in the process. Museum guide should not only have to care for the exposition of the set of artifacts present in the museum, but it has to bring attention also to the way the visitors are perceiving and living the tour.

The complexity of the tasks required to the Guide, and consequently the educational success of the visit, depends on the fact that the communication process is not linear but circular. The objective of the museum guide work is not only to furnish a message or a strict and static information; the guide has to draw educational experiences taking care of the context and stimulating the active participations of the visitors during the museum tour.

The list of tasks (**Table 1**) that the museum guide needs to accomplish during its work can be summarized in three main categories: public speaking, planning, and organizational.

The importance of the dialogue with the museum guide, then, plays a crucial role in the relationship between the visitor and the museum, in particular, for which concern public speaking task. Recently, however, this role was challenged by the competition with digital tools, which usually get a high emotional impact, at the expenses of the cultural value of the experience. Among those devices, robots played in recent years a crucial role. Applications of robots for museum purposes can be resumed in three categories: as a museum guide, in telepresence, like installation.

From a design evaluation conducted on several cases of use, it is possible to say that nowadays, this kind of application presents some deficiencies regarding adaptability to the tour of visit, aesthetic coherence, and easiness of communication with the visitors (in particular robot as museum guide).

From these considerations, the robot Virgil, a tool for museum's guides to enhance their cultural storytelling, has been developed. In the first stage of the project, the robot is driven into the fragile areas of the museum (closed for restoration, or safety) by the museum guide. Visitors could see through the camera set on the top of the robot environments otherwise inaccessible (**Figure 4**).

The museum guide earns benefits in this service because of increase with the robot its communication capability with the visitors. The robot provides the use of a series of multimedia contents such sounds and video, developed by the museum guide and helpful to enhance the visit experience.

Public speaking tasks	Describe tour points of interest to group members and respond to questions
	Provide directions and other pertinent information to visitors
	Escort individuals or groups on cruises, sightseeing tours, or through places of interest
	Monitor visitors' activities to ensure compliance with establishment or tour regulations and safety practices
	Speak foreign languages to communicate with foreign visitors
Planning tasks	Conduct educational activities for school children
	Select travel routes and sites to be visited based on knowledge of specific areas
	Research various topics, including site history, environmental conditions, and clients' skills and abilities to plan appropriate expeditions, instruction, and commentary
Organizational tasks	Greet and register visitors, and issue any required identification badges or safety devices
	Assemble and check the required supplies and equipment before departure
	Train other guides and volunteers

Table 1. The tasks of the museum guide.

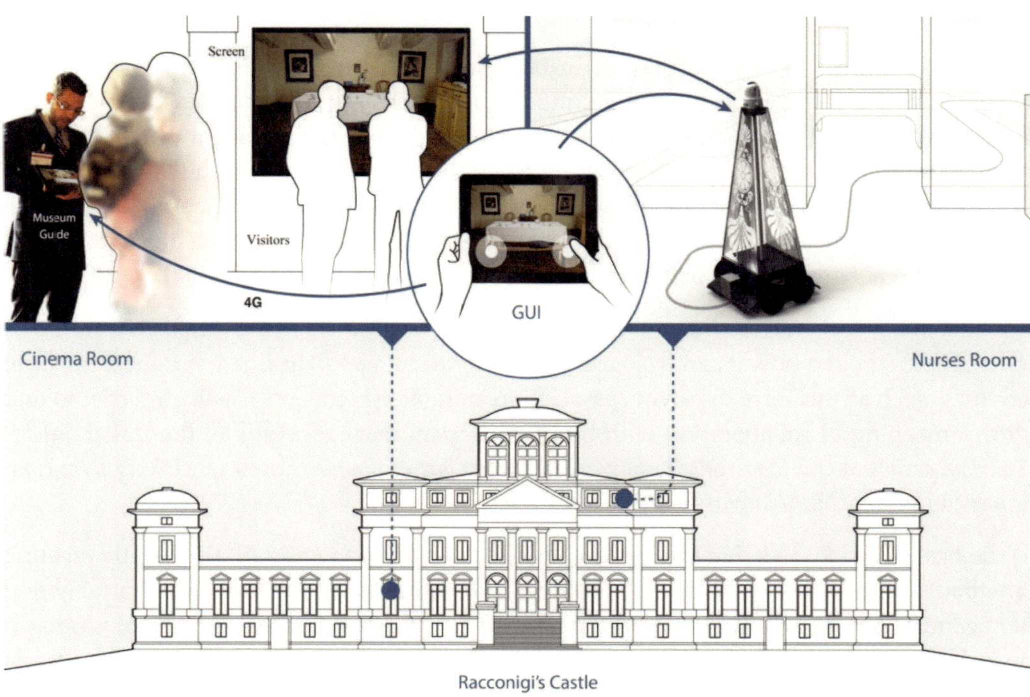

Figure 4. Virgil service concept.

The museum guides assume a central role in the visiting experience because they are entrusted with both the cultural storytelling and the robot control. Human-Robot Collaboration generates an enhancement of human work and further professionalization. Different from the industrial field, the use of robots generated a solid replacement of human work and, consequently, an increase of unemployment [2]. The introduction of robotics in other fields enhances the concern that the same phenomenon could occur. For this reason, during the design process, it is necessary to think about the human work, avoid its replacement and, moreover, enhance it [39].

5. Lessons learned as knowledge for the cultural heritage field

In order to investigate the expertize in the cultural heritage field, it is essential to explore how to attract visitors to their local heritage stimulating the economic and socio-cultural development in a sustainable approach [40]. The development of a local economic growth can be certainly connected to the innovation through the enhancement of new tourism implications, which implies the establishment of new promotional and cultural experiences [41].

Nevertheless, it's possible to hypothesize that if the cultural and social activities are not carefully planned under the aim of a neat social innovation, it's possible to incur "into the decline of heritage sites and an increment of environmental, social and cultural costs" [42]. But speaking specifically about the collection's display inside museums some of the issues emerging from our finding is relate precisely to the alienation of the visitors to the artifacts, which can be translated into a great loss in terms of the display's fruition [43].

Within this context, a roboethic approach can become extremely relevant in terms of sustainability, exploring the economic benefits with impacts on the hosting communities, heritage assets and the environment factors. All these elements will contribute to implement the local wellbeing and cultural promotion of the communities through the use of ICT devices, guided with the help of a service design methodology that follows the guidelines of a roboethic approach in order to enhance the museum experience. The roboethics guidelines help the museum experience enhancement as a tool to better understand and approach the artifacts [44].

In this chapter, we described also the key role of the innovation in the small communities' heritage tourism and how it can welcome the effective change of destination to host the local activities, such as the Terre deiSavoia association, and the Racconigi's Castle. In order to and allow a meaningful collaboration with heritage tourism managers and all the stakeholders. *"Tourism activities and their contribution can be particularly valuable in accomplish long-term commitment to sustainability intents"* [42].

To the best of our knowledge this research wills to contribute to and fill the gap by creating a methodological framework for community participation in heritage tourism planning and management and above all exploring the guidelines for a sustainable and ethical approach toward service robotics applied to cultural Heritage. The research follows a case-study approach and it is currently focused on the Royal Residence of Racconigi.

Nowadays, cultural heritage can be seen as a good testing ground to implement new digital and non digital solution for enhance the user, but most of all, the visitor experience, and

http://dx.doi.org/10.5772/intechopen.69768

heritage tourism, according to Hampton [45] is defined by the will of the visitors to get in touch with their local heritage, and all the historic landscapes, the archaeological sites, local architectures and uses and customs from the past can be a source to feed this will of experience a specific heritage.

Heritage tourism can be also seen as a tool to promote the economical growth of small museums, especially as a bond that connects different expertise from different field of studies or labor [46]. Although not exclusively, heritage tourism has become particularly relevant to culturally rich and remote regions that wish to stimulate growth and compensate for their depressed primary and secondary industry sectors [47].

According to Dragouni and Fouseki [42], creating new connections for a prosperous heritage tourism innovation system should take into account a multi-stakeholder approach, which will define the venue for service design strategies that will take into account social equity and environmental quality, in a sustainable path [42].

These new connections can take advantage of new needs of enhancement of local museums, which could be the starting point of a new academic, social and economical discussion between the stakeholders.

Is possible, therefore, that the value of the roboethics reflections, and above all the main methodological guidelines linked to give importance to the role of people inside the ICT industry, will help to avoid the technological unemployment. On the opposite these reflections will enhance the human factors with a technological help. Prior to enhance the human factors it became essential to give access to people to digital and robotic tools that can improve daily life and labor. In our case study at the Royal residence of Racconigi, was possible to observe the enhancement of the museum guide tools, especially in the human inaccessible contexts. The access to tools and new context would create a positive cycle of cultural heritage valorization in terms of sustainability.

In the Italian Cultural heritage, scenario is possible to investigate and find many local heritages that need the benefit of a well-conducted promotional and dissemination project. These projects could be a positive example of put in evidence how inaccessible places.

The challenge in the described case study was to promote the dialogue between the engineering, the design, the museum, and the academic field in order to enhance the cultural heritage dissemination [48]. One of the advisable outcomes of this research was to create an enveloping design structure that connects professional to museum visitors, giving new hints for experience their heritage.

New connections and museum fruition tools will be born from new needs of enhancement of local museums, which could be the starting point of a new academic, social, and economical discussion between the stakeholders.

It is important to bear in mind the possible bias in response to these new needs, focalizing on how to facilitate the access of visitors to their heritage. Building up a new technology that fosters public spaces to the audience with a roboethic approach is essential in order to make a concrete chance to generate a more accessible culture. Giving tools to understand the heritage can be a turning point into an innovation cultural process, because otherwise the public would not be able to understand the cultural meaning of the artifacts [49].

6. Conclusion

Given the fact that the spectrum of robot's typologies that will be present in our future is constantly increasing, it is becoming crucial to reflect on the ethical implications of human-robot coexistence. From the widely addressed theme of human replacement, especially regarding jobs, it is now emerging the need for understanding where and how the use of robotics is acceptable and desirable. But from primary reflections focused on safety issues, namely how to prevent robot by hurting or replacing people, it is now getting of a primary interest to define way not only to guarantee safety but also to enhance human activities and expertise.

In this regard, design research methodologies can contribute through thanks to their traditional tendency to address projects by simultaneously investigating artifacts, stakeholders, and context. These tree aspects, in fact, are constantly addressed with the attempt of understanding the socio-cultural context and the possible implications that a project could have. For this reason, design research might play a key role in shifting from technology-driven process that currently characterizes robotics to more ethical and acceptable approaches.

As an example of what might entail a roboethic design approach, this article presents the Virgil case study, a telepresence robot for the remote exploration of inaccessible areas of a castle. This resulted from the identification of main ethical design guidelines, and the analysis of both the stakeholders and the context. On one hand, it appeared necessary to empower and enhance the role of museum guides, while on the other hand emerged the issue of inaccessibility for a significant number of areas of the castle.

Providing an answer to these issues represented not only an ethically acceptable solution but also a great chance to innovate and raise the attractiveness of the heritage. So, designing with a roboethic approach is not only a way to do the right thing and avoid undesired drawbacks, it rather represent a desirable way to create new values and opportunities.

Author details

Luca Giuliano*, Maria Luce Lupetti, Sara Khan and Claudio Germak

*Address all correspondence to: giuliano.luca@hotmail.it

Department of Architecture and Design, Politecnico di Torino, Turin, Italy

References

[1] Łichocki P, Billard A, Kahn PH. The ethical landscape of robotics. IEEE Robotics & Automation Magazine. 2011;**18**(1):39-50

[2] Salvini P, Laschi C, Dario P. Design for acceptability: Improving robots' coexistence in human society. International Journal of Social Robotics. 2010;**2**(4):451-460

[3] Rumberger RW. High technology and job loss. Technology in Society. 1984;**6**(4):263-284

[4] Noro K, Okada Y. Robotization and human factors. Ergonomics. 1983;**26**(10):985-1000

[5] Chui M, Manyika J, Miremadi M. Where machines could replace humans—and where they can't (yet). McKinsey Quarterly. 2016 Jul;7

[6] Jamieson S. Robots could replace humans within the next few hundred years, Astronomer Royal predicts. The Telegraph. April 2017. Available from: http://www.telegraph.co.uk/news/2017/04/03/robots-could-replace-humans-within-next-hundred-years-astronomer/

[7] Wakefield J. Intelligent machines: The jobs robots will steal first. bbc.com. September 2015. Available from: http://www.bbc.com/news/technology-33327659

[8] Eden AH, et al. Singularity hypotheses: An overview. In: Singularity Hypotheses. Berlin, Heidelberg: Springer; 2012. pp. 1-12

[9] Simbula C. Professione Robot: 31 lavori che le macchine faranno al posto tuo [in Italian]. Milan: Informant; 2015

[10] Brynjolfsson E, McAfee A. The second machine age: Work, progress, and prosperity in a time of brilliant technologies. New York: WW Norton & Company; 2014

[11] Frey CB, Osborne MA. The Future of Employment: How Susceptible are Jobs to Computerisation. Technological Forecasting and Social Change. 2017 Jan 31;**114**:254-80. ISSN: 0040-1625

[12] Veruggio G. The euron roboethics roadmap. In: Humanoid Robots, 2006 6th IEEE-RAS International Conference on Humanoid Robots. Genova: IEEE 2006 Dec 4 (pp. 612-617).

[13] Muehlhauser L, Helm L. The singularity and machine ethics. In: Singularity Hypotheses 2012 (pp. 101-126). DOI: 10.1007/978-3-642-32560-1

[14] Leonhard G. Technology vs Humanity. London: Fast Future Publishing; 2016

[15] Monopoli A. Roboetica. Etica applicata alla robotica. 2008. [in Italian]

[16] Casiddu N. Human Centered Robotic Design. Firenze: Alinea; 2011

[17] Neumeier M. Metaskills: Five Talents for the Robotic Age. New Riders; 2012

[18] Davies A, Fidler D, Gorbis M. Future Work Skills 2020. Institute for the Future for University of Phoenix Research Institute, 540; 2011

[19] Ford M. Rise of the Robots: Technology and the Threat of a Jobless Future. New York: Basic Books; 2015

[20] Luria M, Hoffman G, Megidish B, Zuckerman O, Park S. Designing Vyo, a robotic Smart Home assistant: Bridging the gap between device and social agent. In: Robot and Human Interactive Communication (RO-MAN). New York: 25th IEEE International Symposium on 2016 Aug 26 (pp. 1019-1025)

[21] Lee MK, et al. The Snackbot: Documenting the design of a robot for long-term human-robot interaction. In: Proceedings of the 4th ACM/IEEE International Conference on Human Robot Interaction; ACM; 2009

[22] Hoffman G, Zuckerman O, Hirschberger G, Luria M, Shani Sherman T. Design and evaluation of a peripheral robotic conversation companion. In: Proceedings of the Tenth Annual ACM/IEEE International Conference on Human-Robot Interaction. Portland: IEEE International Symposium on 2015 Mar 2 (pp. 3-10)

[23] Šabanović S, Reeder S, Kechavarzi B. Designing robots in the wild: In situ prototype evaluation for a break management robot. Journal of Human-Robot Interaction. 2014;**3**(1):70-88

[24] Vandevelde C, Wyffels F, Vanderborght B, Saldien J. Do-It-Yourself Design for Social Robots: An Open-Source Hardware Platform to Encourage Innovation. IEEE Robotics & Automation Magazine. 2017 Mar;**24**(1):86-94

[25] Frayling C. Research in Art and Design. London: Royal College of Art. 1993

[26] Zimmerman J, Forlizzi J. Research through design in HCI. In: Ways of Knowing in HCI. New York: Springer; 2014. pp. 167-189

[27] Freeman E. Strategic Management: A Stakeholder Approach. Boston: Pitman; 1984. ISBN: 0-273-01913-9

[28] Mathur VN, Price AD, Austin SA, Moobela C. Defining, identifying and mapping stakeholders in the assessment of urban sustainability. In: SUE-MoT Conference 2007. Glasgow: Loughborough University, 2007

[29] Bryson JM. What to do when stakeholders matter: Stakeholder identification and analysis techniques. Public Management Review. 2004;**6**(1):21-53

[30] Choi J-H, Lee J-Y, Han J-H. Comparison of cultural acceptability for educational robots between Europe and Korea. Journal of Information Processing Systems. 2008;**4**(3):97-102

[31] Sabelli AM, Kanda T, Hagita N. A conversational robot in an elderly care center: an ethnographic study. In: Human-Robot Interaction (HRI), 2011 6th ACM/IEEE. Lausanne: IEEE International Conference on 2011 Mar 8 (pp. 37-44).

[32] Šabanović S. Robots in society, society in robots. International Journal of Social Robotics. 2010;**2**(4):439-450

[33] Stappers PJ, Sanders EBN. Generative tools for context mapping: Tuning the tools. In: Design and Emotion. London : CRC Press, 2003

[34] Rosson MB, Carroll JM. Scenario based design. Human-computer interaction. Boca Raton, FL. 2009 Mar 2:145-62

[35] Germak C, et al. Robots and cultural heritage: New museum experiences. Journal of Science and Technology of the Arts. 2015;**7**(2):47-57

[36] Elmer A, Gutridge T, Kruse G. Robotic Tour Guide Assistant. In: Proceedings of the 2013 ASEE North-Central Section Conference. Columbus: American Society for Engineering Education 2013

[37] Marcus K. Norio, the robot guide of the Oiron Castle, in tourmag.com. 2014. Available from: http://www.tourmag.com/Norio-the-robot-guide-of-the-Oiron-Castle_a71190. html#

[38] Jonassen D, Land S, editors. Theoretical Foundations of Learning Environments. Abingdon-on-Thames: Routledge; 2012

[39] Bisol B, Carnevale A, Lucivero F. Diritti umani, valori e nuove tecnologie. Il caso dell'etica della robotica in Europa. In: Metodo. International Studies in Phenomenology and Philosophy. Milan: Associazione Metodo 2014 Jan 2;**2**(1). ISSN 2281-9177

[40] Falk JH, Koran JJ, Dierking LD, Dreblow L. Predicting visitor behavior. Curator: The Museum Journal. 1985;**28**(4):249-258

[41] Getz, Donald, and Stephen J. Page. Event studies: theory, research, and policy for planned events. London; New York: Routledge, 2016

[42] Dragouni M, Fouseki K. Drivers of community participation in heritage tourism planning: An empirical investigation. Journal of Heritage Tourism. 2017. DOI: 10.1080/1743873X.2017.1310214

[43] Cimoli, Anna Chiara. Musei effimeri allestimenti di mostre in Italia 1949-1963. Milano: Il Saggiatore, 2007

[44] Steinert S. The five robots—A taxonomy for roboethics. International Journal of Social Robotics. 2014;**6**(2):249-260

[45] Hampton MP. Heritage, local communities and economic development. Annals of Tourism Research. 2005;**32**(3):735-759

[46] Richards, Greg, and Emma, Travis. Cultural Tourism: Global and Local Perspectives. Binghamton, NY: The Haworth Press Incorporated, 2007

[47] Smith L. Uses of Heritage. Oxford: Routledge; 2006

[48] Peressut, Luca Basso. Il museo moderno: architettura e museografia da Auguste Perret a Louis I. Kahn. Milano: Lybra Immagine, 2006. [in Italian]

[49] Heinich N. Il Centre Pompidou e il suo pubblico. Limiti di un luogo utopistico. In: Lumley R, editor. L'industria del museo. Nuovi contenuti, gestione, consumo di massa. Genova: Costa & Nolan; 1988. pp. 226-232. [in Italian]

Section 5

Socioeconomic Impacts

Chapter 7

Mechanical Empathy Seems Too Risky. Will Policymakers Transcend Inertia and Choose for Robot Care? The World Needs It

Johan F. Hoorn

Additional information is available at the end of the chapter

http://dx.doi.org/10.5772/intechopen.70019

Abstract

An ageing population, increasing longevity and below-replacement fertility increase the care burden worldwide. This comes with age-related diseases such as Alzheimer disease and other dementias, cardiovascular disorders, cancer and—hardly noticed—pandemic loneliness. The burden, both emotionally and economically, starts to become astronomical and cannot be carried by those few who need to combine care with work and family. Social solidarity programmes are part of the answer, but they do not relieve the human helper. Yet, many hands are needed where but a few are available. Capacity issues can be solved by the introduction of care robots. Research shows that state-of-the-art technology is such that care robots can become nonthreatening social entities and be accepted and appreciated by the lonesome. Massive employment of such devices is impeded, however, sufficient governmental support of R&D is lacking—financially and regulatorily. This is where policymakers should step in and get over their moral prejudices and those of their voters and stop being afraid of losing political backing. They will regain it in the long run.

Keywords: global ageing, healthcare system, social robots, innovation policy

1. Introduction

It is no news that we have a global ageing problem. For years now, the World Health Organization, the United Nations and national governments publish world-population statistics that show contracting demographic pyramids. Current world population is over 7 billion people. Projections for 2050 are between 8.3 and 10.9 billion, depending on low or high fertility assumptions ([1], p. 2). The Population Reference Bureau expects [2] that by 2050, earth will inhabit about 1.5 billion of people of 65 or older (16% of the world population), whereas in the 1950s, this was a mere 5%. The United Nations expects 32% of 60+ world citizens by 2050 ([1], p. 4, bullet 5).

Not only that, the number of elderly already is higher than the number of younger people (under age 15), and around 2050, estimates are that there will be twice as many elderly as children ([1], p. 4, bullet 5). Below-replacement fertility is persistent for about 30 years now ([1], p. 5, bullet 14), and one can see that in China, for instance, one-child policy together with a growing economy will lead to many poor Chinese parents taken care of by a few prosperous children. Yet, more elderly that are taken care of by a minority of the young are a direct threat to economic growth, worldwide. To counter such adverse effects, Germany plans to increase the influx of immigrants with imaginably, more integration problems. For the elderly, top priority is to counter loneliness, and care managers also state that self-management is prime.

The number of single-person households is increasing, and people are getting older than before ([1], p. 6, bullet 19). Moreover, people expect low-cost 24 h high-quality care. A considerable number of older people suffer from loneliness and social isolation, which negatively affects their health and well-being [3]. Loneliness is an epidemic in modern society [4]. Increasing longevity not only adds to an ageing population feeling lonely but also introduces specific health issues, such as increasing cardiovascular diseases and cancer. In their joint 2011 publication, the National Institute on Aging, National Institutes of Health and World Health Organization estimated that globally, between 27 and 36 million people suffer from Alzheimer disease or other dementias ([5], p. 14). After age 65, the occurrence of dementias doubles with every 5 years of age (ibid.). In addition, annual new cases of cancer will reach 17 million by 2020 and then continue to grow up to 27 million by 2030 ([5], p. 19). Long-term care is required and will be centred on 'home nursing, community care and assisted living, residential care, and long-stay hospitals' ([5], p. 23; also, see Ref. [6]).

In view of what lies ahead of us, it is not only the question how many people we can recruit from a declining population to take care of the old and frail but also how many healthy people will be left to keep the economy going that covers the formidable expenses involved? How will 'beanpole families' ([5], p. 22) make sure that enough informal care is supplied to keep their parents 'healthy longer, delaying or avoiding disability and dependence' (ibid., p. 23) and who is going to earn all that money to cover the expenses?

The good news is that apart from social solidarity programmes, which still require the effort of many, there are technological innovations almost ripe enough to help make up for the loss in care support. By now, evidence accumulates that robots can for a large part compensate social isolation and sustain self-management, much to the relief of the elderly (e.g. [7–9]). They can assist (e.g. the Riba II Care Support Robot for Lifting Patients); they can monitor (e.g. the NEC PaPeRo, Partner-type Personal Robot, for telecommunication), and they can provide companionship (e.g. AIST Paro, a therapeutic robot seal).

A number of social robots are available on the market and already successful. Nao/Zora does physical exercises, brain games and dance moves and can lead the bingo. Paro, the baby seal, squeaks, cries and can recognize its given name and responds to touch. It can reach deep-dementia patients where humans have lost all contact. Kaspar is used for autistic children to practice with emotion expression and ways of conduct. Zeno and his sister Alice R50 work with the elderly; and then of course, there are robot surgery arms, drawer pullers and lid openers, but they do not aim for companionship and bonding.

2. Problem statement

Almost everyone in the care supply chain is in favour of employing social robots, each for their own reasons. Managers see cost reduction; care professionals see a decrease in work pressure (although they fear dehumanization). Although a little ashamed [10], informal caretakers admittedly find it a relief, and clients are happy being helped [11]. The laggards are the top decision-makers in national politics, in multinational corporations and the board of directors of insurance companies and housing cooperatives. Their contribution would lie in financial and regulatory aid to make novel laboratory findings available to the public at large. Instead, they constantly worry if they will lose the support of their voters, shareholders and allies, and hence, they freeze. At face value, mechanical empathy seems politically too risky.

3. Purpose of the chapter

On behalf of the old and lonely and their loved ones, this chapter wants to mobilize policymakers to take action and get involved to deal with a global situation that is rapidly deteriorating. The two things they have to do are first to provide *just enough financial leeway* to develop laboratory prototypes into devices that are ready to market, facilitating the take-up by service providers. The other thing is the *reduction of rule pressure* so that innovations can be easily tried and tested without worrying too much about ornate grant proposals, detailed managerial accounts, intellectual property rights, codes of medical ethics, etc. Those demands may punch in once the robots are market-ready.

The Japanese have understood this better. In 2013, the Tokyo Times announced 25.3 million dollars in governmental subsidies to stimulate a knowledge-driven care-robotics industry [12]. Transparency Market Research (cited in The Economist) expected in 2014 that the market for medical and care robots will rise from US$ 5.5 billion in 2011 to US$ 13.6 billion in 2018 [13]; so where is the rest of the world? They stand by and watch.

4. Social action

Those who have a social network currently mobilize their family and friends to organize their future care. This is done in small networks where professionals and informal caretakers negotiate care tasks and so release the formal care networks (e.g. [14]). To enable people to combine care with work, it is suggested that organizations for home care should assign professionals to extra-residential caretakers of single-household clients [15]. If intercultural mediation is required, social workers may be engaged (cf. [16]). In spite of the good intent, however, the care burden may be so heavy that work and care cannot be combined effectively [17].

Of course, the call for solidarity is genuine, and that people should help each other goes without saying. But apart from this moral stance, it should be possible, we should be able to, and we should have the capacity to deliver. Will there be enough hands, and can we afford those hands, a very banal but very practical matter. So where robot help is feared because of humanitarian considerations, humanitarian considerations are at stake if we cannot help everyone equally well, and people become emotionally and physically impoverished. Robots are feared because of job loss, but in the long run, there will be insufficient people for those jobs anyways, so we better make sure robots are up to speed to fill in for us. Voters may demand the creation of more jobs, but what are they good for if soon there will be no one to fulfil them?

5. Low-hanging fruit

Luckily, little is needed technology-wise to make a positive contribution to the health and well-being of the socially impaired. For example, research indicates that a feeling of social presence and relatedness is already established if the robot simply invites a response [18], is modest and polite [19] and has a kind of cuteness to it (e.g. [20]). That is not too hard to establish. Toy factories have lots of experience (enter Bandai's Tamagotchis). The only thing toy makers do not have is knowledge of what to program, how to keep the interaction going, and what health protocols to follow (exit Tamagotchis).

However, there is quite some concern whether empathy should be left to a machine. Point is that human care is not always that empathic either. Day care starts in the morning (half an hour), in the afternoon the nurse looks whether the client is alright (15 min) and in the evening, the client is helped with dinner (another 15 min). The other 7 h of the day is filled with staring out of the window and counting the busses that drive by. Loneliness can be so aggravating that anything will do—even a friendship machine.

We conducted a social experiment with robot Alice on the couch of three grandmothers [21]. This experiment was recorded in a by now worldwide known documentary *Alice cares*, directed by Sander Burger in 2014 [22]. The experiment was to find out if a human can build up a relationship with a machine that pretends to be empathic. If the documentary (and also our other research) makes one thing clear, it is that this is so (which is since the Tamagotchi craze perhaps not surprising). We know better now what to look for when building a social application. Good speech is more important than serving coffee. Automatic conversation analysis is more important than gait technology. Face recognition does far more than hand gestures. In addition, the robot should not talk too much about itself but invite the users to share their feelings. Above all, the robot itself must be placed in a position of dependence and assist the needy in an unobtrusive manner [21].

Is this different for men and women? We do not have hard results here, but observations led us to believe that men are equally interested as women, but they take a detour. Women directly start to bond with the machine, whereas men first test it and want to know how the technology works, before they make friends with it.

One might counter that robots are rational and cannot work with emotions. Point is that robots do not necessarily have to have an empathic understanding of humans in order to raise empathy. People love their cats and dogs and attribute all kinds of human traits to them. Likewise, when the robot falls of a chair or is 'hurt' in any way, people feel sorry for the machine. In one of our studies [23], we observed that people reckoned that after maltreatment, the robot must feel sad (!).

How important is it, then, to formally model emotion processes and try to simulate empathy with a machine? Scientifically, this is very important. In trying to make a formal model, the scientist stumbles upon his/her own implicit assumptions and finds many research gaps. When the model works fine (is inherently consistent) and in simulation, participants cannot distinguish its behaviour from that of humans (is empirically valid), we may argue that we understand human nature better than before. You only understand it if you can build it.

In application, proper emotion models are less important. Users find other functionalities (e.g. chatting about the weather) far more interesting, also in companion machines. Emotion is a nice to have, not a must have. If some emotional stimulus comes from the robot, that will do the trick. People will then project and attribute all kinds of human-like qualities to the machine and deem it 'emotionally responsive'. They do not hold it for a human, but the responses are recognizable enough to evoke a human response, like an animal does. The way that is achieved, through a full-fledged psychology-based artificial intelligence (AI) or some artistic-engineering trickeries, is indifferent. They take the bait anyways.

6. What a social robot is capable of, now and in the future

The Alice R50 we work with is designed by David Hanson from Texas and manufactured by the firm Robokind. The machine we have right now is reassembled from the working parts of two machines that collapsed most of the time, and the basic software that drives the motors, speech engine, and cameras was completely reprogrammed by Germans Media from Amsterdam. Apart from that, we developed software programs to steer Alice's and other robots' behaviours, for example, to simulate friendship with the user [24].

In the documentary film *Alice cares*, you see Alice in action on the sofa of three elderly ladies. In the beginning, the ladies were invited to our office (well-controlled environment), talking to a robot that followed our conversational version of the MANSA protocol for care evaluation, using closed questions (also well controlled).[1] That led to very grumpy replies, but yes, here the robot acted autonomously most of the time. In the apartment of the ladies, however, we were in open environments, and the conversations went from coffee to photo books to football to singing and children and back again, so here we had Alice in remote control and she was helped quite a bit by the engineer. The only things she did autonomously in this situation were following faces, blinking, head motion, lip-sync and translating speech input to text and typed-in text back to speech.

[1]http://isp.sagepub.com/content/45/1/7.short?rss=1&ssource=mfc

At the moment, we are in the process of making better conversational software. In view of current state of the art speech technology, natural language processing, conversation analysis and knowledge representation in the semantic web, it should be feasible to have a simple well-defined chat about the coffee, the weather, the family, the money machine, an appointment with the doctor and so forth. But that does take some more development and technology integration work.

Thus, if you want to talk about 'a social robot', you have to distinguish between the robot as hardware (the puppet) and the robot as software (machine behaviour). Within the software, there are three things it should be doing: perceiving (data input), processing (information throughput) and executing actions such as speech and facial expressions (output).

In addition, you should distinguish between well-controlled environments such as lab settings and open environments such as a living room. The software that potentially can drive a robot's behaviour is really good at throughput in well-controlled environments (e.g. [24]).

What robot software is really bad at is perception and execution, which is limited to following a face (input) or standard speech recognition and production and some head movements, smiles and frowns (output). Yes, there are better systems available (e.g. the Google Assistant), but common researchers have no access to them (e.g. IBM Watson—the full version, that is). No one is allowed to see the source code, and private data are free for taking. Moreover, most robot hardwares are engineered pretty badly and, after intensive use, show quite some wear and tear.

In other words, state-of-the-art social robots could make better use of the AI that is already available such as computer vision and deep learning. On the output side, there are hardly sufficient expression possibilities: robots should have better facial expressions, gestures, conversational phrases, etc. However, that is not AI but interaction design. Many AI modules for learning, reasoning, etc. are in a testing phase and sometimes seem to be working pretty well. However, the numbers those systems produce may indicate that the system 'likes its user', but that is not translated yet into communication that is fit for human consumption (speech, facial expression, etc.). So the robots one sees on the Internet are partly a mock-up, partly prototype and partly an autonomous system. That the robot is operated in remote control and is not independent is also due to the fact that the software engineering of all the different components is not done right yet: too many loose ends and badly integrated systems and too unreliable for grandma on the couch. But we are progressing.

If all this is ready, we may have a system that can independently speak and have some understanding of its user and the situation both are in. Unavoidably, people will think of feature movies in which robots take over the world and in which advanced AI outsmarts humanity; and also serious people such as Hawking and Musk add fuel to the flames by declaring that AI may be the biggest threat to humanity ever. But why would you build a car without a break or an airplane without landing gear? You do not have to learn a robot everything we know. You do not have to give it the capacity to learn evil things on its own. AI does not grow from a tree. Humans create it. My standpoint is that humans are the biggest threat to humanity and in their pursuit of supremacy, some nasty people produce nuclear bombs, war gasses and, yes, also killerbots. But they do not have to. Yet, we allow them to. We use anything as a weapon, anything. That is a humanistic and not a technological problem. It is politics in its essence.

7. People respond differently to social robots than to smartphones or fellow humans

Certain critical thinkers do not get what the fuzz is all about. If a social robot is merely a plastic hull around an ordinary computer that has some fancy AI, then why bother about the childish appearance? You could have everything installed on your smartphone, and it would do the same. You could even use an animated avatar to make it more human-like, but actually even that is pure nonsense. What is the difference between this technology and, for instance, the use of monitors in the context of telecare, apart from the visual image of a robot?

All this is true, but the visual image of the robot is the quintessence. Critical thinkers look at technology way too rational and functional. This is psychology. The back side of the temporal cortex of babies of a few days old already responds to faces, not to smartphones. In confrontation with a social robot, the human emotional system takes precedence over cool cognition as soon as the relevance of an unfulfilled need—here companionship—is higher than the question whether that need is met by a human being or a machine. This is why a robot—physically present, eyes, voice—is more effective than a monitoring camera, an app on a smartphone or a 2D avatar on a screen. People tend to forget that there may be a human operator behind the machine—and here we touch on shaky ethical ground obviously. The robot is just human enough to invite self-disclosure of the user (cf. [25]) but socially peripheral enough that a person may not expect any consequences to the life confessions one may do. The response to the robot is stronger and more personal than to an avatar on a screen or a monitoring camera. Hence, people take the role of grandparent over the robot, feel the need to feed it and look straight into the camera-eyes. The response also is more personal than to the average human being because there are no social repercussions to what you say or do. Hence, revelations and family histories are told that no one ever heard of.

8. Security: you lose self-disclosure when outsiders start snooping

Social robots invite users to disclose personal information—even things they do not tell to other human beings. That is a powerful quality for therapeutic purposes; and if the robot is connected to a care centre, this information can help good-willed caretakers understand how to help or treat a person. But it also makes privacy violations all too easy for those snooping around: managers, insurers, governments, tech companies, tabloids and the police.

Suppose that a social robot is monitoring a patient, autonomously, and equipped to call the ward when someone falls (fall detection). The first problem is that of the missed signals and false alarms. How does a robot know that someone really has tripped over? There are so many situations in which a person may hit the floor: for an incidental exercise, playing with the cat, making fun with the grandchildren, just fooling around and simulating a fall to draw attention. What exactly are the physical signals that indicate 'emergency' and how to make sense of the context in which those physical indications occurred? Either we have a robot that constantly warns for nothing (false alarms) or that is so 'contextually sensitive' that it overlooks the crucial symptoms (missing the signal). This makes the robot unreliable, and even when something *is* the matter, the nurses at the ward may stumble into 'the pitfall of the liar', not believing the robot when they actually should. Put the robot back into remote control then. Have a person look over its shoulder. But then we are also back to spying around, gossiping or, worse, not telling the patient that s/he is observed.

Robots nowadays are often implemented based on remote servers, for instance, for converting speech to text. That may actually violate people's privacy [26]. The server can easily keep a permanent record of what people have said and make it available years later to companies or governments that might mean those people no good (ibid.). This is bad news if robots are to care for millions of people in a decade or two. The goal is to stop pervasive poking and prying.

The only solution would be to convert the audio to speech locally in the robot's on-board computer, never transmitting it elsewhere [26]. The goal is to make sure the robot does not send the speech data anywhere, not as audio and not as text. Rejecting speech-to-text translator services is a prerequisite for this. However, just rejecting those services is not enough. We need to be sure the robot does not in fact send anyone that data. For the same reasons, the software in the robot should be free software. Otherwise, the next 'upgrade' might include surveillance—as in so many other products (ibid.). All this does not necessarily mean to disconnect the robot from the Internet. All computing can be done inside the robot, and the Internet connection is used merely to receive and send other data when the user wants to communicate with someone else. The robot must not be designed to farm out its own computational activities and function to some other computer or remote service. To a limited extent, this may reduce the potential usefulness of the robot, but 'freedom sometimes requires a sacrifice', and usefulness is not the only virtue to strive for (Ref. [26]).

For the robot not to be someone else's snoop, it must not send the user's commands (as audio or as text) to anywhere else. In practice, why would it send the audio to somewhere else? For speech-to-text conversion. Therefore, to avoid sending the audio to somewhere else, the robot must do speech-to-text conversion locally. Doing the conversion locally is necessary, but not sufficient, for respecting privacy. The robot could send the audio somewhere else, even after doing the conversation locally, but it must not do so [26].

Of course, as soon as there is some sort of transmission of information over Wi-Fi (e.g. between the server in the closet and the robot on the couch), the locally stored information is open to interception again. Even worse, one could plug a USB stick into the robot's back and download personal information manually. Luckily, there are some protections against such special, targeted surveillance. They are legal protections: for the state to do this, it needs a court order. For others to do, it is a crime. Probably, few people will be the target of a search warrant. Most likely no one will burglarize someone's home and steal personal data, but you never know (Ref. [26]).

The great danger is from social robots and other systems that systematically surveil everyone (Google, Facebook, Skype, Amazon). That is what we have to fight. Social robots could be such systems, or we could design them not to be (Ref. [26]).

http://dx.doi.org/10.5772/intechopen.70019

9. Moral machines

We could design a robot that refuses to reveal personal information to those who are not authorized by the end user. Police systems also have such graded access. Some information is accessible to the street cops, other merely to the detectives. Whether robots make a positive or negative contribution to society depends on how we humans deal with them. For example, if people learn robots nothing about weaponry, they will not know anything about weapons. But there are always people who still want to send a robot to war and teach it how to use weapons (e.g. Russia's Fedor). How robots act depends purely on what they learn from us or what we allow them to learn by themselves.

Instead, a robot also could be used as a moral compass for the biased decisions people make. A robot can reason from first principles and is not led by personal emotions. If a medical specialist was forced to choose between saving your child or five others, I know what you choose. The moral robot would not: 5 is more than 1. Your kid dies. Game over [27].

10. No cold motives in a declining industry

Managers will fire personnel, that is, cold-hearted managers will, because greed is their main motive. If life fulfillment was their driver, managers would train their staff to work with robots that support people in their jobs as a relief of tedious and undesired work, leaving the enjoyable work to human employees. Suddenly, workers work more efficiently, enthusiastically and as a result effectively. Productivity increases per capita and so does the manager's annual profit share.

Uninformed directors will decide on questionable grounds to encourage the use of technology in healthcare and other service professions, and indeed, directors are notoriously uninformed; and once they are informed, they become notoriously hesitant. They tend to follow bad policy if everybody else does rather than be the forerunner of good policy. That is why directors are not leaders.

But after management sacked the outmoded workforce, they will find out they have to hire a different type of professional. Although the unions are on strike—the rally against robots—they have to learn that if you do not adapt, you die out. Professions vanish primarily because their practitioners fail to adapt to changing circumstances. Clog making and lace bobbing have become folklore. Others will jump at the opportunity and fill up the niches in the professional market. Before the twentieth century, nobody ever thought that air pilot could be a profession. Who would ever have thought that 'ethical hacking' is a profession today? Soon we need plenty of robot programmers, interaction designers, conversation analysts, robot liability lawyers and engineers in the medical domain, education, hospitality, etc. and not before long, HRM will be short for human-robot relationship manager.

Care is costly, financially and emotionally; and we soon will have emptied both resources. Robots may relieve both workforce and clients. There is one warning, however: if robots actually make true what they potentially promise and indeed reduce care pressure, this should by no means lead to dim-witted managerial cutbacks and layoffs. Certainly, cost reduction is much wanted but only if the quality of care is guaranteed if not enhanced. Robots have their unique selling points, but that inevitably means that humans have other qualities that together bring care to the next level, such that happier people are healthier people and, thus, cost less. If robots are capable of keeping elderly people 6 months longer at home, we have a business case in place.

11. What makes a robot special (and so, what a human)?

If programmed well, a social robot invites self-disclosure (cf. [25]). People confess to strangers things they will never do to their family or caretakers. Such confessions may not only disclose important medical information (e.g. not taking medication); expressing oneself has therapeutic value of its own.

With sufficient security measures installed, social robots guarantee the privacy of information. They do not gossip or go behind your back. There are no social repercussions to what you say or do. Social robots will not laugh at you if your mouth is sore, and you do not wear your dentures. They have patience, patience if you are slow in responding and patience in reminding you of an appointment, to do exercises. They have excellent memories. They can play memory games with you.

If designed well, their appearance and behaviours are nonthreatening and child-like. That also gives room for asking silly questions (fun), asking things it already knows (training your memory) and asking impolite questions (evoking self-disclosure), like a grandchild.

Social robots may invite social and physical activation, not for you but to humour them. So going out to see the neighbour is not because you are lonely but because the robot wants to have a chat. And finally, social robots are in no need of claiming any social space. You can complain to them for hours on end, and they do not feel that attention should be equally distributed among the conversation partners or that you bore them. On the contrary, if they bore you, you simply turn them off.

A smartphone or tablet will not do the trick. Devices for the elderly have to have a physique, they should have eyes and a voice and then they are ready to deliver whatever functionality in a highly appreciated manner (cf. [28], p. 96).

While robots take care of the mundane repetitive tasks (e.g. chit-chat, calendar keeping, exercise coaching); humans can do the sophisticated work: if someone has a stroke or needs a true shoulder to cry on. But because the human caretakers did not burn their energy on the tiresome day-to-day chores, they are happy to provide full attention and devotion in cases of emergency or psychological despair.

12. Research programme

The documentary *Alice cares* [22] recorded the pinnacle of a research programme that investigated the direct effects of robot companionship on the loneliness of elderly. However, we also conducted more indirect studies, such as the study I did with Eva N. Wijker, looking into how elderly of the future ($N = 128$, between 50 and 65, $M_{age} = 56.6$, $SD = 4.5$, 73% female) would feel about social robots at the age of 85, both in a hedonistic sense (i.e. companionship) and a utilitarian sense (i.e. as exercise coach). Half of the respondents received the depiction of a senior woman conversing with a Nao/Zora robot; the other half saw the same woman physically exercise with Zora. An online structured questionnaire with Likert-type items and rating scales (1 = totally disagree, 6 = totally agree) then inquired about the robot's good intent (scale ethics, $M = 5.00$, Cronbach's $\alpha = 0.77$), its action possibilities (scale affordances, $M = 4.12$, $\alpha = 0.92$), how important the robot was to the respondents' goals and concerns (relevance, $M = 3.75$, $\alpha = 0.91$), their intentions to use the robot ($M = 3.39$, Spearman $\rho = 0.83$), how emotionally involved ($M = 1.95$, $\alpha = 0.93$) they were with the robot and in how far they felt at an emotional distance ($M = 4.08$, $\alpha = 0.90$). Divergent validity of the measurement scales was confirmed by principal component analysis (oblique oblimin rotation).

We ran a one-way MANOVA with function (companion vs. exercise coach) as between-subjects factor and ethics and affordances as within dependent variables. Effects of function on ethics were not significant ($F < 1$), but on affordances they were ($F_{(1,117)} = 5.95$, $p = 0.016$), indicating that the functionality of the robot as exercise coach was rated higher than the robot as companion. Another One-way MANOVA with function (companion vs. exercise coach) as between-subjects factor had use intentions, involvement and distance as within dependents. Pillai's Trace indicated significant effects of function on the dependents ($V = 0.13$, $F_{(3, 124)} = 6.10$, $p = 0.001$), while contrast analyses showed that exercising evoked significantly more involvement ($M = 2.23$, $SD = 1.62$) and less distance ($M = 3.74$, $SD = 1.28$) than companionship (involvement: $M = 1.62$, $SD = 0.68$, $p = 0.000$; distance: $M = 4.49$, $SD = 1.14$, $p = 0.001$). There were no significant differences in use intentions ($p = 0.065$). Covariate analysis showed an effect of gender on distance ($V = 0.03$, $F_{(3,118)} = 1.19$, $p = 0.014$): women felt more emotional distance towards the robot than did men ($F_{(1,120)} = 4.34$, $p = 0.039$). Thus, robot Zora as an exercise coach elicited more warm feelings (involvement) and less detached feelings (distance) than as a companion robot, whereas intentions to use the machine did not differ beyond coincidence.

Next we performed regression analyses, particularly Hayes' mediation-moderation analyses, to find out across both functions how relevant ethics and affordances would be for use intentions, feeling involved with the robot or at a distance. Across functions, ethics ($b = 16$, $p = 0.541$) did not influence relevance, but affordances did ($b = 0.53$, $p = 0.000$). With relevance as a mediator, affordances were indirectly effective for use intentions ($b = 0.68$; $p = 0.000$), involvement ($b = 0.43$, $p = 0.000$) and distance ($b = -0.60$, $p = 0.000$). Additionally, affordances also had a direct effect on involvement ($b = 0.25$, $p < 0.01$) and distance ($b = -0.48$, $p = 0.000$), even without touching upon the goals and concerns of the respondents. Scrutiny of the individual effects showed that indeed affordances were directly influential for involvement (bootstrap $b = 0.25$, $p = 0.001$, [0.142, 0.354]), distance (linear $b = -0.48$, $p = 0.000$) and use intentions (linear $b = 0.54$, $p = 0.000$). Thus, the more action possibilities were perceived in the Zora robot (whether exercising or conversing), the higher the use intentions, the higher emotional involvement was with the robot and the least distance was felt.

According to the respondents, the function of companionship showed fewer action possibilities (affordances) than exercising. Therefore, we explored in how far affordances even for the function of companionship would explain relevance, use intentions, involvement and distance. Regression analysis according to Hayes showed a significant effect of affordances on relevance ($b = 0.44$, $p = 0.000$) as well as a direct effect on involvement ($b = 0.22$, $p = 0.001$) and distance ($b = -0.40$, $p = 0.000$). Furthermore, affordances were directly influential for use intentions ($b = 0.22$, $p = 0.001$). Affordances explained 15% ($R^2 = 0.15$) of the variance in use intentions, 19% of involvement ($R^2 = 0.19$) and 22% of distance ($R^2 = 0.22$). In repeating these analyses for the function of exercising, surprisingly, affordances explained less variance in use intentions (11%) and distance (16%) than for the function of companionship. For exercising, affordances had no direct effect on involvement with the robot at all.

We concluded that elderly of the future at face value see more possibilities in exercising with a robot than in companionship. Ethical concerns were not an issue for both functions, so we need not worry about that in future implementations. These elderly of the future thought overall that Zora meant it very well ($M_{Ethics} = 5$). They had neutral to moderate intentions to use the machine ($M_{UseInt} = 3.39$), which did not differ between exercising and company and deemed the robot somewhat relevant ($M_{Relevance} = 3.75$).

However, they had more friendly feelings (involvement) and felt less aloof (distance) when the robot was exercising than when it was deliberately used as an antidote to loneliness. In other words, telling elderly of the future they may become lonely and for that they will receive a robot may to them not be a pretty outlook. But the detour through an exercise coach that by the way also counters loneliness by building up a friendship might just do the trick.

In general, if action possibilities (affordances) touch upon important goals and concerns (relevance), friendly feelings (involvement) and wanting to use a robot (use intentions) increase, whereas feelings of distance drop. This we also found for the few action possibilities respondents saw in the companion robot. Remarkably, however, respondents also thought that even if companionship-affordances were not relevant to them personally, they still showed direct effects on feeling involved with the machine and feeling less at a distance, and most strikingly, they were more willing to yet use the machine: Seeing is believing. By keeping the robot out of the sphere of personal relevance, people already get used to the idea of having one for companionship later.

Of course, the respondents in this study were still socially active; many of them had (part-time) jobs. In due time, however, when family and friends become fewer, loneliness may strike out. Therefore, to know about the mitigating effects of social robots on loneliness in the long run, we plan on a new line of research. Yet, a major hurdle is to conduct laboratory experiments with seniors. The experiments are usually tiresome, the elders often too frail. Therefore, much of what we know does not come from systematic research on truly old people. To tackle this problem, a research programme is wanted that bases itself on a firm methodology that can be used to answer a range of questions and variables with a common

http://dx.doi.org/10.5772/intechopen.70019

line of reasoning. The backbone of that programme should be twofold: the use of *indirect* but nevertheless *structured observations* (1) in a *switching replications design* (2) in which control and robot-treatment groups swap. This approach goes beyond observations and case studies, exploiting a hybrid methodology in which observation is combined with quasi-experimental field studies.

We need two groups of seniors, one of which is split in two so there are three groups in total. Group 1 is a group that claims not to have any problems. Suppose our dependent variable is 'loneliness'; then these people are diagnosed as not feeling lonely. Groups 2a and 2b do feel lonely, but the moment in time that they receive a robot companion differs.

Measurement happens at three points in time (so-called waves) to see how long-lasting the effect of robot intervention is. Participants are observed in their room through video surveillance for 1 h. This material will be cut back to ± 15 min. Then a trained interviewer will pose a number of open questions like 'So, how is life; how are you doing?' (± 10 min). The interview also is videotaped, but the interviewer is not visible on camera.

For the seniors in Group 1, the 1 hour interview pre-period is spent without robot support; for Group 2, the pre-period is spent with robot support. Participants in a session do their robot-guided treatment as they please. The advantage is that seniors can stay in their room; they do not come to the laboratory because the robot will be brought to their homes. Each person is videotaped during robot interaction (this is the interview pre-period) and during the interview.

The cut back video footages (about 25 min) of the period preceding the interview (the 'pre-period') and the interview itself are the stimulus materials that are viewed by a large number of observers to assess the state of mind of the senior participant, using a structured questionnaire (e.g. about loneliness, well-being, physical condition, etc.).

For senior Group 1, little has to be arranged as compared to Group 2, the treatment group, which interacts with and via a robot intervention. As said, Group 2 is split into two and enters a switching replications design: a before-after repeated-measures design with and without, in this case, robot help. The dependent measure(s) could be anything (e.g. loneliness, quality of life), but what we want to know is whether robot intervention works or not and how durable the effects are. Note, however, that all kinds of variables of interest can be measured and analyzed this way. Here is an example with well-being as dependent measure:

Group 1	Well-being before (t1)	No robot	Well-being after (t2)	No robot	Sustained well-being? (t3)
Group 2a	Well-being before (t1)	Robot	Well-being after (t2)	No robot	Sustained well-being? (t3)
Group 2b	Well-being before (t1)	No robot	Well-being after (t2)	Robot	Sustained well-being? (t3)

As Group 1 apparently has no problems, we have no predictions for the outcome variables. Or it should be that the level of well-being remains about constant at all time points ($t1 \approx t2 \approx t3$). It is more of a research question (RQ) rather than a hypothesis what the difference may be with Group 2a and 2b.

More specifically, Group 2a tests whether robot interference (the intervention or the treatment) has a durable effect. Group 2b tests whether having no robot works just as fine or that the robot yet has an effect even after being in crisis for a longer period of time.

It is *in*essential that the main effect of group (2a vs. 2b) at t3 is significant. After all, the groups 2a and 2b could end up equal balance, showing the same net results. Yet, the interaction between group (2a vs. 2b) and time (t1 vs. t2 vs. t3) of measuring well-being is crucial and should show the following pattern: in Group 2a, two outcomes may underscore the robot-inspired-well-being hypothesis The first is the *lift-off-level-out* effect, where well-being at $t1 < t2 \approx t3$. The second is the *lift-off-boost-up* effect, where well-being at $t1 < t2 < t3$. The second effect indicates that people in Group 2a were inspired and learned the skills to fire up their own well-being without extra help of the robot. In Group 2b, the robot-inspired-well-being hypothesis is sustained when we find $t1 \approx t2 < t3$. All other outcomes will count as refutation of long-term beneficial effects of robots.

13. Finance

It takes parents about 20 years to programme a child. We call this 'upbringing' or 'education'; and after that, people are still telling each other how to behave, what to do and what not (cf. culture, legislation, ethics). Moreover, it is about 4 million years ago that the first prototype humanoid Lucy was developed, found in Ethiopia, and it took the last 125,000 years to come up with a model that is now the benchmark against which we compare our social robots: today's modern human beings.

By contrast, and if I take it broadly, social robotics is developing for about 20 years, and programming our humanoid Alice took us about 4 years, which is 1/5th of the time to raise a modern-human child.

Modern humans, then, as pinnacle of their superior intelligence, are prepared to put up 1.9 billion Euros to make clearer pictures of the stars beyond our galaxy, wonderful pictures indeed. They look like abstract modern art. But running costs of the space telescope Hubble are yet another 5.6 billion Euros plus five space shuttle missions for maintenance and overhaul. That is a sum total of about 10 billion Euros for just one machine.

The investment plan that we submitted to the national science council to create a better version of the Alice machine amounts to the sum of 3.8 million (that is a fraction of 1 billion), including VAT and including a business plan to produce 200 machines in the post-grant period that can be readily applied to hospitals, care centres and elder homes. Moreover, everything will be open source so that a complete new industry can arise based on this one investment, impacting education, coaching, hospitality, entertainment, etc.

In other words, modern man is prepared to invest maximally to solve a minimal problem (i.e. clearer pictures) and minimally to solve a maximal problem (i.e. global ageing).

In 25 years' time, there will be two elderly in need of care against one younger caretaker. The WHO and UN foresee a worldwide trend. So are we going to educate more care professionals? Yes. Or bring in more foreigners to do it for us? Yes. But both measures do not solve

http://dx.doi.org/10.5772/intechopen.70019

the lack of sufficient hands, globally. They merely relocate the problem. Moreover, we need firefighters, a police department and school teachers as well. But to counter the hardship of an ageing world population and install a completely new industrial sector—the 3.8 million Euros to refurbish an old device will not do.

Although there is a lot of positive feedback on what social robots can achieve, and we are told to 'keep up the good work', nobody puts their money where their mouth is. Society misses a sense of urgency of facing a mass of old people taken care of by a handful of young. It is not just eldercare. The economy will collapse under an unproductive and overstrained population in which those who can work have to take care of their parents and grandparents. Robots not only fulfil tasks in eldercare but also in all other walks of life, freeing up the time of those who are the motor of our financial system, society and welfare.

As a result, relatively small research groups and development teams work in isolation on specialist robot functionality. Because those machines can do one thing very well but all other things not, people are not impressed. Since they are not impressed, there is no financial support to develop one robot that can do many things very well. Without integration of the many machines doing one thing well, we are not talking about a robust solution to serious societal issues. Without empowerment, we leave potential world saviours work at the level of ardent hobbyists. As human race, we thought it necessary to develop a nuclear bomb. Let us turn that zeal into something more positive and organize a Manhattan Project for social robotics and develop the Ultimate Android [29].

Social robots like Alice can be used to help mild healthcare cases, so that the heavier cases can be catered for by people. One person in a home with light care (sheltered housing, little help) costs around € 2000—a month. If we keep 200 people out of the elder home for 6 months (or 400 people for 3 months), we save € 2.4 million. For € 2.4 million, we can design the blueprint of a robot that can keep 200 people stay at their homes for 6 months longer. That € 2.4 million is a *cost-effective* investment, the only way to cut costs *responsibly*. With all the robots that the industry produces based on that blueprint, society makes nothing but profit, which can be invested into long-term care, performed by human hands. It is quite a detour to say that in about 40–50 years, we simply do not have enough hands to take care of everyone, but if politics and business invest now, we make that care cheaper and better than today. People and robots work together, and they specialize in what they are good at. The main barrier is those first few millions.

14. Conclusions

The global population is ageing, and the number of people that can provide care diminishes. This brings moral disgruntlement and shame, which are unwarranted because nobody can take care of a dozen of people and at the same time have a job, run a business, run a family and have friends. Without drastic measures, today's shaky economic recovery will be undone. New social participation programmes in which people work part-time may be one solution. However, that will lower gross national income and puts many into a position of fulfilling an obligatory job instead of being fulfilled. Not everybody is fit to or finds satisfaction in helping others, all of the time. That will not enhance but reduce the quality of care.

In 2025, we will be with 8,141,661,000 people inhabiting the earth of which 32% will be under 20, which is a drop back of 8% compared with today. In less than 10 years' time, relatively small younger generation has to take care of 800 million people over 65, a growth of 410 million elderly worldwide [6]. This increase is not evenly spread across the globe. In Latin America and Asia, the increase in elderly people will hit 300% (ibid.). This actually is happy news because it means that globally people live healthier and, therefore, longer. Perhaps, we just need to reproduce at a higher rate.

The planet's population exceeds the 8 billion in 2025, and if we stop reproducing, and demographics shows we do, there will be insufficient people to take care of the older generations. But if we increase our 'replacement fertility', we empty the planet's resources. And minimizing our medical and health programmes so that fewer survive beyond childhood is no option either. Instead, the trend is the provision of low-cost 24h high-quality healthcare.

The alternative is a technological solution that fills in the gaps when needed and can be disposed of once redundant, a workforce you can increase on demand and lay off just as easily, because they do not care. Technological innovations are progressing, and the basic requirements that elderly have on such systems can be provided for. The only hurdle is that technology such as social robotics is not robust and fine-tuned enough to release them to the market. All the stakeholders in the care supply chain—if done sensibly—underwrite the potential benefits of robot care except for the CEOs, the ministers and the chair of the board of directors. They wait and hold their cards close to their chests, where a warm heart should be pounding.

If employed properly, care robots bring cost savings. They also initiate a new industry. This new industry will not only be care related; innovations will radiate to other areas such as hospitality, education, sports, coaching and therapy. The only things needed are easy-access money (fuel) and regulatory respite (grease), typically things policymakers deal with for a living. Policymakers, it is your turn. Fuel and grease that robot engine because we are ready to fire up!

15. Suggestions

Seven suggestions to society (one a day) that can be encouraged by governments of organizations and nations in setting an example:

- Leave moral outrage and fear of job loss behind.
- Focus on the opportunities (both in health and economy).
- Cost savings through health improvement, not by massive layoffs.
- Provide just enough easy-access money for R&D.
- Reduce the regulatory burden.
- Start designing new business models.
- Be brave.

http://dx.doi.org/10.5772/intechopen.70019

Author details

Johan F. Hoorn

Address all correspondence to: j.f.hoorn@vu.nl

Faculty of Social Sciences, Vrije Universiteit, Amsterdam, The Netherlands

References

[1] United Nations, Department of Economic and Social Affairs, Population Division, New York. World Population Prospects: The 2012 Revision, Key Findings and Advance Tables [working paper no. ESA/P/WP.227] [Internet]. 2013. Available from: http://esa.un.org/wpp/Documentation/pdf/WPP2012_%20KEY%20FINDINGS.pdf [Accessed: 13-02-2017]

[2] Haub C. Population Reference Bureau, Washington DC. World Population Aging: Clocks Illustrate Growth in Population Under Age 5 and Over Age 65 [Internet]. 2011. Available from: http://www.prb.org/Publications/Articles/2011/agingpopulationclocks.aspx [Accessed: 12-02-2017]

[3] Sparrow R, Sparrow L. In the hands of machines? The future of aged care. Mind and Machine. 2006;**16**(2):141-161. DOI: 10.1007/s11023-006-9030-6

[4] Killeen C. Loneliness: An epidemic in modern society. Journal of Advanced Nursing. 1998;**28**(4):762-770

[5] Suzman R, Beard J. National Institute on Aging, National Institutes of Health, World Health Organization. Global Health and Ageing [NIH Publication 11-7737] [Internet]. October 2011. Available from: https://www.nia.nih.gov/sites/default/files/global_health_and_aging.pdf [Accessed: 12-02-2017]

[6] Beard J, Officer A, Cassels A, et al. World Health Organization, Geneva, SW. World Report on Ageing and Health [Internet]. 2015. Available from: http://apps.who.int/iris/bitstream/10665/186463/1/9789240694811_eng.pdf [Accessed: 13-02-2017]

[7] Banks MR, Willoughby LM, Banks WA. Animal-assisted therapy and loneliness in nursing homes: Use of robotic versus living dogs. Journal of American Medical Directors Association. 2008;**9**(3):173-177. DOI: 10.1016/j.jamda.2007.11.007

[8] Stafford RQ, McDonald BA, Li X, Broadbent E. Older people's prior robot attitudes influence evaluations of a conversational robot. International Journal of Social Robotics. 2014;**6**(2):281-291. DOI: 10.1007/s12369-013-0224-9

[9] Sharkey A. Robots and human dignity: A consideration of the effects of robot care on the dignity of older people. Ethics and Information Technology. 2014;**16**(1):63-75. DOI: 10.1007/s10676-014-9338-5

[10] Broadbent E, Tamagawa R, Patience A, Knock B, Kerse N, MacDonald BA. Research attitudes towards health-care robots in a retirement village. Australasian Journal on Ageing. 2012;**31**(2):115-120. DOI: 10.1111/j.1741-6612.2011.00551.x

[11] Zsiga K, Edelmayer G, Rumeau P, Péter O, Toth A, Fazekas G. Home care robot for socially supporting the elderly: Focus group studies in three European countries to screen user attitudes and requirements. International Journal of Rehabilitation Research. 2013;**36**(4):375-378. DOI: 10.1097/MRR.0b013e3283643d26

[12] The Tokyo Times. Japan Develops Robot Industry for the Elderly Care [Internet]. 2013. Available from: www.tokyotimes.com/japan-develops-robot-industry-for-the-elderly-care/ [Accessed: 13-02-2017]

[13] The Economist. Robot Nurses Are Big Business. From Surgery to Companionship, Japan Looks to Robots for Healthcare [Internet]. August 1, 2014. Available from: http://geloo-kahead.economist.com/robot-nurses-big-business/#sthash.pmNaTAyl.dpuf [Accessed: 13-02-2017]

[14] Broese van Groenou MI, Jacobs MT, Zwart-Olde NE, Deeg DJH. Mixed care networks of community-dwelling older adults with physical health impairments in the Netherlands. Health and Social Care in the Community. 2016;**24**(1):95-104. DOI: 10.1111/hsc.12199

[15] Plaisier I, Broese van Groenou MI, Keuzenkamp S. Combining work and informal care: The importance of caring organisations. Human Resource Management Journal. 2015;**25**(2):267-280. DOI: 10.1111/1748-8583.12048

[16] Parast SM, Allaii B. The role of social work in healthcare system. Journal of Social Science for Policy Implications. 2014;**2**(2):59-68

[17] Jacobs MT, Broese van Groenou MI, De Boer AH, Deeg DJH. Individual determinants of task division in older adults' mixed care networks. Health and Social Care in the Community. 2014;**22**(1):57-66. DOI: 10.1111/hsc.12061

[18] Hoffman G, Birnbaum GE, Vanunu K, Sass O, Reis HT. Robot responsiveness to human disclosure affects social impression and appeal. In: Sagerer G, Imai M, Belpaeme T, Thomaz A, editors. Proceedings of the 2014 ACM/IEEE International Conference on Human-Robot Interaction (HRI '14); 2014; New York. NY: ACM; 2014. pp. 1-8. DOI: 10.1145/2559636.2559660

[19] En LQ, Lan SS. Applying politeness maxims in social robotics polite dialogue. In: Yanco H, Steinfeld A, editors. Proceedings of the 2012 ACM/IEEE International Conference on Human-Robot Interaction (HRI '12); 2012; New York. NY: ACM; 2012. pp. 189-190

[20] Moyle W, Cooke M, Beattie E, Jones C, Klein B, et al. Exploring the effect of companion robots on emotional expression in older adults with dementia: A pilot randomized controlled trial. Journal of Gerontological Nursing. 2013;**39**(5):46-53

[21] Hoorn JF, Konijn EA, Germans DM, Burger S, Munneke A. The in-between machine: The unique value proposition of a robot or why we are modelling the wrong things. In:

Loiseau S, Filipe J, Duval B, Van den Herik J, editors. Proceedings of the 7th International Conference on Agents and Artificial Intelligence (ICAART); Jan. 10-12, 2015; Lisbon, Portugal. Lisbon, PT: ScitePress; 2015. pp. 464-469

[22] Doolaard J, Burger S. Alice Cares [Documentary]. The Netherlands: KeyDocs; 2014. Available from: https://vimeo.com/116760085 [Accessed: 12-02-2017]

[23] Konijn EA, Hoorn JF. Empathy with and projecting feelings onto robots from schemas about humans. International Journal of Psychology. 2016;**51**(S1):837

[24] Hoorn JF, Pontier MA, Siddiqui GF. Coppélius' concoction: Similarity and complementarity among three affect-related agent models. Cognitive Systems Research. 2012;**15-16**:33-49. DOI: 10.1016/j.cogsys.2011.04.001

[25] Bhakta R, Savin-Baden M, Tombs G. Sharing secrets with robots? In: Viteli J, Leikomaa M, editors. Proceedings of EdMedia: World Conference on Educational Media and Technology; 2014; Waynesville. NC: Association for the Advancement of Computing in Education (AACE); 2014. pp. 2295-2301

[26] Stallman R. Personal Communication. January 2-25, 2016

[27] Pontier MA, Widdershoven G, Hoorn JF. Moral Coppélia—Combining ratio with affect in ethical reasoning. In: Pavón J, et al, editors. Lecture Notes in Artificial Intelligence. 7637th ed. Berlin-Heidelberg, GE: Springer; 2012. pp. 442-451

[28] Tanaka K, Nakanishi H, Ishiguro H. Comparing video, avatar, and robot mediated communication: Pros and cons of embodiment. In: Yuizono T, Zurita G, Baloian N, Inoue T, Ogata H, editors. Collaboration Technologies and Social Computing. Communications in Computer and Information Science. 460th ed. Berlin, Heidelberg, GE: Springer; 2014. pp. 96-110

[29] Hoorn JF. The Ultimate Android [Internet]. February 11, 2016. Available from: http://www.robopop.nl/wp-content/uploads/2016/09/2016-02-11-JFHoorn-The-Ultimate-Android.pdf [Accessed: 15-02-2017]

Chapter 8

Electronic Prescribing and Robotic Dispensing: The Impact of Integrating Together on Practice and Professionalism

Roderick J. Beard

Additional information is available at the end of the chapter

http://dx.doi.org/10.5772/intechopen.69769

Abstract

Technology developments offer hospital pharmacies opportunities to enhance efficiency and safety in the dispensing process. Adoption of technology potentially allows release of resources to develop more patient-focused activities. Resource release can be achieved via a variety of impacts, such as efficiency of the dispensing process, reduction of potential for dispensing errors and potential to adjust skill mix. Developing more patient-focused activities can enhance pharmacy development in a broader sense, and as this happens, changes can occur in professional identities across a range of job roles within the pharmacy. These changes offer benefits to the development of the professional model.

Keywords: electronic prescribing, robotic dispensing, efficiencies, safety, professionalism

1. Introduction

Anderson [1] describes a succinct history of pharmacy, charting it from early periods to the present day. He describes how the Sumerians in 2000 BC were using plant drugs and during the reign of King Hammurabi (1795–1750 BC), a system of laws, known as Hammurabi's Code, was established. Medical and surgical practices were regulated, and diagnosis and treatment were separated from the preparation of medicine, which were carried out by assistants or apothecaries. A primitive understanding of drugs enabled formulations of draughts, mixtures, infusions decoctions, medicinal wines, ointments and poultices. Some are still in use today (myrrh, poppy, thyme, liquorice, peppermint, cannabis etc.). Around 1500 BC, the Egyptians had advanced practice beyond the Assyrians. In Egyptian society, diagnosing and treatment of illness was separated from making medicines, though pharmacy was regarded as a special branch of medicine. In Egyptian mythology, Thoth, the scribe of the gods, was known as ph-ar-imki (translated as 'warrant of security'). This represents the origin of the word pharmacy.

In ancient Greece, the word 'pharmakon' was used for a process of cleansing or removing bad luck or circumstances. The pharmakon was often the person who was blamed for the ill luck and either cast out from the city or worse. However, the word was used in the context of people being made to feel better by a pharmakon. The key point is that for the last 4000 years across many cultures, while techniques and medicines have progressed, the considerations of professionalism, patient safety and use of latest developments are as relevant today as they have ever been.

2. Electronic prescribing and robotic dispensing

In 2005, the UK Department of Health issued a report authored by the Chief Pharmacist 'Building a safer NHS for patients: improving medication safety' [2]. This was a detailed paper on medication errors, the causes and potential remedies, and represented a development by the Department of Health from the paper 'An organisation with a memory' [3]. The paper 'Building a safer NHS for patients' made many suggestions on how errors could be minimised by designing them out through the use of a system's approach to medication systems. In the paper, the use of electronic prescribing (EP) and robotic dispensing (RD) was put forward as potential tools to help reduce dispensing errors. The problem is that the use of electronic prescribing and robots is not systematically documented in the literature as to what features provide the greatest safety. There is a variety of design in EP and robotic dispensing systems, and it is important when surveying the literature to consider the context of the medication system in a hospital. For example, literature from the USA has a different pharmacy model in contrast to the UK, [4] where payment for services is through insurance and not state funded. This means in the USA, tracking cost for re-charge to insurance companies is an important component of activity, and unit dose dispensing is much more of a feature of US hospital pharmacies. This is not the case in the UK, which needs to be remembered when looking at published literature.

Karsh [5] commented that generally, health care was poor at implementing new technology to patient benefit. In the UK, over 400 (2010) dispensing robots have been installed (source ARX Ltd.), but the literature does not identify all the potential benefits the robots may give. The installation of so many machines in hospitals implies that buyers not only envisage benefits but also efficiencies in the dispensing process, since there is also potential benefit with regards to skill mix adjustment in the dispensary.

Electronic prescribing (EP) is a subject that has produced many papers in the literature, but many of the studies originate in the USA and reflect to some extent the American hospital pharmacy model. Medication errors are a primary focus of reports in the literature with little focus on other aspects and benefits of EP. There are fewer publications on robotic dispensing, and a paucity of information on combining both EP and robotic dispensing, which would be of a wider interest to pharmacists generally.

http://dx.doi.org/10.5772/intechopen.69769

3. Benefits of electronic prescribing (EP) and robotic dispensing (RD)

When considering the possible benefits of combining an EP system and robotic dispensing together, certain criteria need to be satisfied to yield the maximum benefits to the pharmacy, patients, to hospital organisations and indirectly to the value of the profession. These are as follows:

- An integrated EP system throughout the hospital to allow flexibility of working points and minimising re-keying of data and making relevant patient data available anywhere in the hospital. In this context, integration means the EP system links to other modules of the hospital systems, such as the patient administration system, the pharmacy system, pathology system, nurse administration system, chemotherapy medication system and so on.
- If a robot is used in the pharmacy, it is better that it is directly linked to the EP system to increase efficiency (avoidance of re-keying). Hospital pharmacies will generate hundreds of thousands, if not a million-plus, of labels a year, and re-keying data (to produce labels) can add up to a significant amount of time. The system also requires automatic labelling to maximise the benefits. Integration across hospital departmental systems also potentially permits use of pathology results to link to drug selection and prescribing. For example, a low white blood cell count may be a useful screen prompt when prescribing methotrexate. However, there needs to be care when considering such types of screen prompts, as information overload to the prescriber will have a tendency to render key messages likely to be unheeded.
- As the above criteria are met, resources will be released internally in the pharmacy to allow working processes to be re-modelled to perhaps focus to a greater degree on ward-based activity.
- The use of EP to do the 'policeman' functions of formulary management means there are opportunities for enhanced professional relationships with nursing and medical staff, where pharmacists can fully utilise their clinical skills. This happens because EP can be operated restrictively, and this can be used to prevent doctors prescribing 'off formulary'. The un-rewarding (and potentially challenging) task of changing what doctors want to prescribe is removed. The pharmacist is no longer a formulary policeman but the professional who helps prescribers navigate the EP system. This positive element to the work helps in ward relations and helps embed pharmacists in ward processes.

Benefits will be harder to realise at the ward level unless the pharmacy department has enough pharmacists with the clinical skills to deliver an enhanced professional role. Well-trained pharmacists are crucial to the whole process, and unless pharmacists are capable of delivering the service medical and nursing colleagues expect, technology implementations will under-deliver on wider benefits.

- Using automation technology does not denigrate the role of more mundane medicine distribution and procurement. If anything, it enhances it if wisely implemented, and skilful use of technology can reduce the number of staff needed to be involved in this activity (within the confines of the pharmacy walls at least).

- By doing more 'higher' professional activities and less mundane work, the level of patient centred expertise of pharmacists and technicians is expanded and enhanced.
- Enhancing the clinical service into the broader aspects of hospital speciality activities increases potential pharmacy identities and potentially adds to the survivability of the pharmacy service. The impact is necessarily not only on pharmacists but also on other pharmacy staff. The potential for a wide range of electronic systems (EP, RD, 'smart' 'drug cupboards', etc.) in the pharmacy means technical and other supporting staff is presented with new opportunities too. The developments in technology allow new roles for these staff in supporting these systems and developing expertise to maintain them. As such, these developments (along with ward-based activities) replace more traditional medicine preparation skills.

It is not an inevitable consequence of automation and deploying technology that an enhanced service will follow. Planning is required to make this happen. Some hospitals could choose to use the efficiency created to reduce the number of pharmacy staff employed. If this was the corporate decision, there would be no enhanced clinical service but that would be the organisation's choice. The technology creates the efficiency, what happens after that depends on competing priorities at a corporate level.

However, should a pharmacy department look to implement EP-RD systems, the type of robot chosen may limit efficiencies, and the integration levels of the EP may also put constraints on efficiencies. The pharmacy management team must also think of the end game of any implementation programme. They must have a vision of what they want once the systems have been built. The pharmacy team needs to have a clear strategy of what the EP-RD system would be used to support and further develop the clinical service, which means that there has to be pharmacists suitably trained to be able to move into roles that the freed time allows.

The 'Building a safer NHS' paper [2] quoted a study from the dispensing error analysis scheme (DEAS) published by Cardiff and Vale NHS Trust [6] and analysed errors from 66 contributing hospitals from 1991 to 2001 and which included 7000 errors. As such, it represents one of the biggest surveys of its kind in the UK. The following categories of errors were recorded by frequency in **Table 1**.

This profile of errors provides a plan with which technology can be adapted to remedy the potential dispensing errors. With regards to the UK, several authors have commented on robots in the UK pharmacies. In 2004, Swanson [7] described all the various types of dispensing machines from the chaotic storage type of ARX machines to the channel fillers produced by such companies as Baxter. Whittlesea and Phillips [8] described the efficiencies expected and the fall in error rates associated with robotic dispensers. No figures for errors were quoted, and the non-robotic dispensing rate (i.e. using staff) was quoted at 10 items per person per hours. James [9] looked at workload in non-robotic and robotic environments. She quoted a non-robotic dispensing rate of 7.25 items per hour and a robotic dispensing rate of 12 items per hour. She also quoted a dispensing error rate of 637 items per 100,000 for the manual system and 338 per 100,000 for the robotic system. The data did not specify if this only referred

http://dx.doi.org/10.5772/intechopen.69769

Type of error	Proportion (%)
Wrong drug	23
Wrong strength of correct drug supplied	23
Wrong quantity	10
Wrong warnings or directions	10
Wrong drug name on the label	9
Wrong strength on label	8
Wrong form	7
Wrong patient name on label	7

Table 1. Frequency and type of dispensing errors catalogues in DEAS study.

to items moved into the robot. Goundrey-Smith [10, 11] quoted a 50% error rate reduction at Wirral Hospital on installing a chaotic storage robot and a 16% error rate reduction with a channel-filler robot at Wolverhampton post installation. However, sometimes in papers, key features of the technology are not described, so it becomes difficult to assess what features are the most important in achieving tangible benefits. Cantrill et al. [12] looked at three types of EP systems in North West England. They surveyed medical and pharmacy staff and concluded on the basis of the surveys that the benefits of EP were difficult to quantify. Reference to the appendix in the paper [12] outlines the functionality of the systems being used in the surveys. It becomes clear that the systems used were not integrated as described in this text and therefore not surprising that enthusiasm for EP across organisations was variable. The paper is a good illustration of understanding how the levels of functionality can affect benefits and perception of benefits. In some cases, some doctors were using dual systems (i.e. a 'Kardex' and also an EP system), so it is not surprising where they are required to do more work, they become less positive about the technology. However, when looking at the potential for technology to assist in reducing dispensing errors, we can consider the DEAS error profile, and see how technology can have an impact on reducing dispensing errors.

The following table identifies the potential error rate reduction potential of integrating an EP and RD system together.

It follows from the analysis of **Table 2** that provided the EP and robotic dispensing are integrated in a specific way, many dispensing errors can be 'designed out' by skilful application of technology.

The key points to remember are as follows:

- Because EP system is integrated, when the doctor prescribes the medicine on the computer, he is also in fact writing the label to attach to the medicine. This means the label is always what the doctor requested.
- Because the label is always accurate to the prescription so there is no labelling error.

- If the RD system is used to store medicines by bar code, then drugs can only be supplied by the robot by bar code identification. There is then a direct electronic link between the medicine, bar code and item selected on the electronic prescription and the label that is printed. This is the most crucial step in deriving safety benefits from the technology. It is achieving this direct link through integration of the technology that delivers safety benefits. To design in these links is to design out potential errors.
- Once so designed, the system works from anywhere in the hospital. This allows significant amounts of dispensing activities to be triggered outside the pharmacy, but only works if there is a direct link between prescription, label and robot.
- Automatic labelling is a critical component of this system.
- Once medication has been checked by a pharmacist in the EP system, it is possible to make the dispensing nearly instantaneous. The remaining bit of the process is to get the medication from pharmacy to the ward promptly.
- In achieving 'instantaneous dispensing', the role of the pharmacist potentially changes. No longer are pharmacists directly in control over the whole dispensing process. It is akin to craftsmen producing goods being replaced by production lines where quality control is through process control, and each individual is responsible for a part of the overall process, not all of it. This has some significant implications for the way the pharmacy is subsequently managed and run.

A skilfully designed system should produce zero errors for the robot plus EP system combined, potentially a huge benefit in safety. This is far better than quoted in the literature for robotic dispensing [8–10] and should be regarded as significant. The system design prevents errors. A key feature of this is an automatic labeller, which means there is no point in the dispensing process for human intervention, and crucially, potential human error. Because of the EP system, no re-keying of data is required, so the process becomes more efficient.

Type of error	Proportion (%)	EP prevents	RD prevents	EP + RD
Wrong drug supplied	23		Y	Y
Wrong strength of correct				
Drug supplied	23		Y	Y
Wrong quantity	10	Y		Y
Wrong warnings or directions	10	Y		Y
Wrong drug name on the label	9	Y		Y
Wrong strength on label	8	Y		Y
Wrong form	7		Y	Y
Wrong patient name on label	7	Y		Y

Table 2. Types of errors prevented by EP and RD system.

Dispensing rates can be significantly faster than quoted in the literature. However, dispensing in the pharmacy is not entirely risk-free, since not all items are supplied and labelled from the robot (for example, oral chemotherapy). Clearly though, the opportunity for errors is significantly reduced.

When compared to the system outlined by Reifsteck et al. [13] and Gonidec et al. [14], the comparisons start to encounter the contextual difficulties of understanding the medication systems in other countries (in these cases, USA and France). Reifsteck describes his hospital as having 900 beds and says his hospital in Albuquerque consistently ranks among the top ten integrated delivery networks in the USA. They use computerised practitioner order entry (called EP in the UK) and integration with a unit dose pharmacy robot. The hospital in the USA also has a closed loop patient administration system. Gonidec et al. [14] describe the use of a robot combinded with electronic prescribing in a French prison (for 700 prisoners). They used a unit dose system similar to quoted by Riefsteck [13]. Gonidec quotes an error rate of 0.5% post implementation, but these are quoted as mainly wrong location delivery of orders triggered by the EP system. Gonidec also quotes a production rate of 377 doses per hour and the types of error and the number of times they occurred in the 3-month study, but they bear no resemblance to the DEAS study and relate to operational problems within the unit dose system.

Reifsteck et al. [13] quote a drop in error rate from 23.5% pre-automation to 9.9% post automation. Their figures include wrong time administration errors (defined as being plus or minus 1 hour of specified time). When administration errors are not included, the rate falls to 1.9%. Nowhere in the document are dispensing errors mentioned. Without wrong-time administration errors, there were 10 errors logged, 7 being wrong administration techniques.

4. Potential methods of evaluating EP-RD

4.1. Turn around time for prescriptions

There is potential within an EP-RD system to increase the speed of turnaround time from the point of the clinical check of the prescription to nearly instantaneous, which is not a common feature within substantial numbers of UK hospitals at present. However, at very busy periods, the dispensing times can potentially rise, depending on the capacity of the RD system chosen. In traditional 'Kardex' hospital systems, dispensing times can often be up to 4 hours for non-urgent dispensing [15]. Beard (no relation to this author) and Wood quote how, by using 'Lean' processes, they reduced the dispensing time of the prescription from 4 h to around 1 h in a pharmacy which had robotic dispensing (these times include the time it takes a signed prescription to get from ward to pharmacy).

4.2. Dispensing rate

Whittlesea and Phillips [8] and James [9] quote benchmarks of around 10 items per person per hour. The linking of EP and the robot means there is an efficiency increase, simply because of avoiding re-keying in data. Dispensing times can have a potentially significant fall. This efficiency does not compromise on safety. The EP component of the system gives in effect a huge digital capacity compared to a traditional Kardex-based system. The robotic component adds significant picking and mechanical capacity and is limited only by the limits of what can be configured in a particular dispensary. The potential dispensing rate is then only limited by the number of picking heads that are deployed.

4.3. Out of hours supply

The automation of supply of products means that anything that is in a robot could be supplied remotely. This feature is used in many hospitals in the UK. A night safe can be built into the pharmacy to allow nursing staff to collect items sent to the night safe chute. The on-call pharmacists could have remote access to the software via a hospital laptop and also have access to the patient's prescription and other relevant data (e.g. pathology) via EP. Supplies, if appropriate, can be made out of hours without the pharmacist attending. While this is an attractive feature, there is a potential downside in that the convenience of such systems creates the potential for work 'drift' from wards from normal dispensing times into times that are intended for emergency or urgent needs only.

4.4. Returned stock

There is a benefit opportunity from robot installation to make it easier to manage stock returned which could be re-issued if an automatic loading hopper is a part of the RD system. The effect is more a staff behavioural one rather than direct effect, but if ward's stock is returned, prior to robot installation, there is work required by the store or dispensary staff to put the items back on the shelf. Even if it is just a few packets of different items, it can require a relatively large amount of work to put these items back on the appropriate shelf. With the robot hopper, it is relatively easier and less time consuming for staff to put these returns on the loading hopper and re-using the products is made easier. All store or dispensary staff have to do is put returned stock on the conveyor, and the robot does the rest. Because it is easy to re-cycle stock, stock is re-cycled and wastage potentially falls. However, the benefits of this aspect depends on where a hospital pharmacy already is with regard to its current stock re-use, but an automated loading hopper should either permit greater increase use of returned stock or allow less staff to be used in this process.

4.5. Extended hours of service

As the EP-RD system increases efficiency, it should facilitate the manning of addition hours easier, as less staff should be needed in the dispensary than would be required in there were no such system. As such, the EP-RD system would reduce the cost of providing an extended hour's service by reducing the manpower needs (in the dispensary at least). Outside normal working hours, staff (covering late nights and weekend shift workers) may not be as familiar with stock control methods and be tempted to take supplies from shelves without recording them. Each individual may think it is only one omission, but the compounded effects over time can thwart the re-ordering algorithms, leading to a tendency to overstock 'just in case'.

Benefit	Impact
Error reduction	Zero within EP-RD system
Dispensing times	Nearly instantaneous
Value of stock returns?	Value returned to be re-issue
Reduction in stock holding?	May be significant
Reduction in dispensary staff?	Can be reduced
Skill mix of dispensary staff?	Possible lower skilled staff (lower staff costs)

Table 3. Summary of benefits from robot implementation linked directly to EP.

By installing a robot, it potentially denies access to the robot shelves, and forces staff to follow standard procedures, making the stock recorded on the computers more accurate.

All of the above identified benefits have a significant impact on the efficiency of the pharmacy. The cumulative effect is to make the dispensary processes more efficient. This frees resources for other work and can create a means for significant departmental developments. These are summarised in **Table 3**.

5. Implications of EP-RD on the professional model

As a result of implementing combined integrated EP and RD systems in the pharmacy, it becomes apparent there are further implications for pharmacy beyond operational efficiencies. The professional model potentially can be changed or developed further, and professional identities for all pharmacy staff can be modified.

One implication of the EP-RD model is the dispensary pharmacist 'oversees' the supply of dispensed medicines leaving the pharmacy, but he may not have seen many of the prescriptions and has not had any sight of the dispensing process. This can be because his pharmacy colleagues are doing the dispensing process controlled from ward level though the EP-RD system. He is faced with a series of labelled packets for patient X and trusting that his colleagues have all done their part at ward level, permits the dispensed medicines to be dispatched to the ward.

5.1. Impact on locums

One consequence of changing the operational model is the training time required to get new members of staff competent to work with EP-RD systems. Some sites report it takes around 3 months for new staff to gain enough knowledge to be confident of using an integrated EP-RD system at a competent level. This has implications for ensuring proper induction of staff when starting, requiring staff available to manage the training process. It also has a potential negative effect, which means that the short-term locum cover is not really possible. Locum pharmacists are very much part of the pharmacy profession, providing cover when the regular pharmacist(s) are not there. In the traditional model, because their expertise is focused on the paper prescription, whichever pharmacy they are in, they can clinically check the prescription and check for accuracy. In the EP-RD model, they have to be trained in the systems to find the electronic prescription and know what functions to use to provide the clinical check. Because the electronic system is highly integrated, all the information a pharmacist needs is available within the computer (pathology tests, nursing notes, radiology reports, clinic bookings, previous admissions, etc.), but the pharmacist needs to know where to find all of this information. The consequence of not being able to use locums means that as a department, the hospital pharmacy has to have a process for replacing staff that leave with a system of timely succession planning.

5.2. Skill mix

The installation of an EP-RD system could trigger efficiencies with regard to skill mix in the dispensary. This is because in an EP-RD system, there is no need for interpretation of the doctors' handwriting (since prescriptions are electronically written); it is possible to de-skill the activity of dispensing and checking, since the professional check now becomes mainly a process check.

5.3. Other parameters

As mentioned above, other parameters of efficiency may become possible such as stock holding, stock turn, dispensing error reduction, speed of dispensing and stock re-cycling. The impact of EP-RD should show improvement. This frees pharmacy staff, both pharmacists and technicians, to be potentially deployed in ward areas and other clinical settings.

6. Taxonomies in other disciples; mapping a taxonomy for hospital pharmacy

It may be worth considering if the EP-RD model changes the professional model of pharmacy? If so, what are the implications for the future? Is it likely to have a positive or negative impact on the profession? Will it increase professionalism, reduce it or be neutral?

Looking at other taxonomical models may give some predictive insights. Bloom [16] laid down his taxonomy of learning. It is described in a hierarchy. Its aim is to summarise different thought processes to create continual sustained learning. This is summarised in **Figure 1**.

Similarly, this hierarchy model has been applied to informatics by Ackoff in the 1980s, when looking at methods to describe the relative value of information [17], as in **Figure 2**.

In Ackoff's model, he carefully defines the words as:

Data = symbols

Information = processed data

Knowledge = applying data and information (how questions)

Understanding = Appreciation of 'why'

Wisdom = evaluated understanding

The significance of the top level is that it represents a better capacity to manage an environment. As Ackoff pointed out, organisations ought to become 'wise' because it confers potential sustainability of the organisation. It permits the better survival of the organisation.

Goundrey-Smith in his text [10] describes a similar 'Bloom' style of hierarchy for electronic prescribing systems and is summarised in **Figure 3**.

The hierarchy Goundrey-Smith describes for electronic prescribing has similarities to Ackoff's model. The assessment for technology and its impact on professionalism could be mapped against pharmacy activities, to provide more details of practical use. If we look at hospital pharmacy activities, we could list them as:

Procurement, distribution, dispensing, ward stock supply, out-patient services aseptic manufacturing, non-sterile manufacturing, medicine information, audit, clinical services, running clinics, clinical trials activities, research work, R&D work, management functions, weekend services, on-call services, attendance to interdisciplinary meetings, finance work, reporting to within organisation, staff induction, staff training, record keeping and controlled drugs management. These are described in **Table 4**.

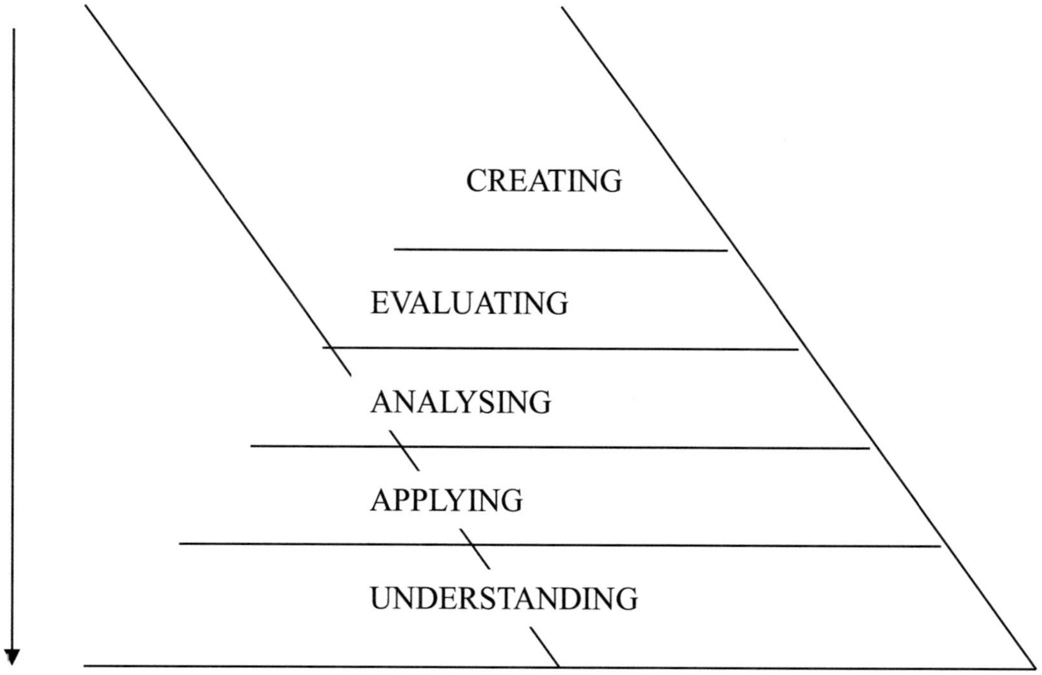

Figure 1. Summarising Blooms taxonomy.

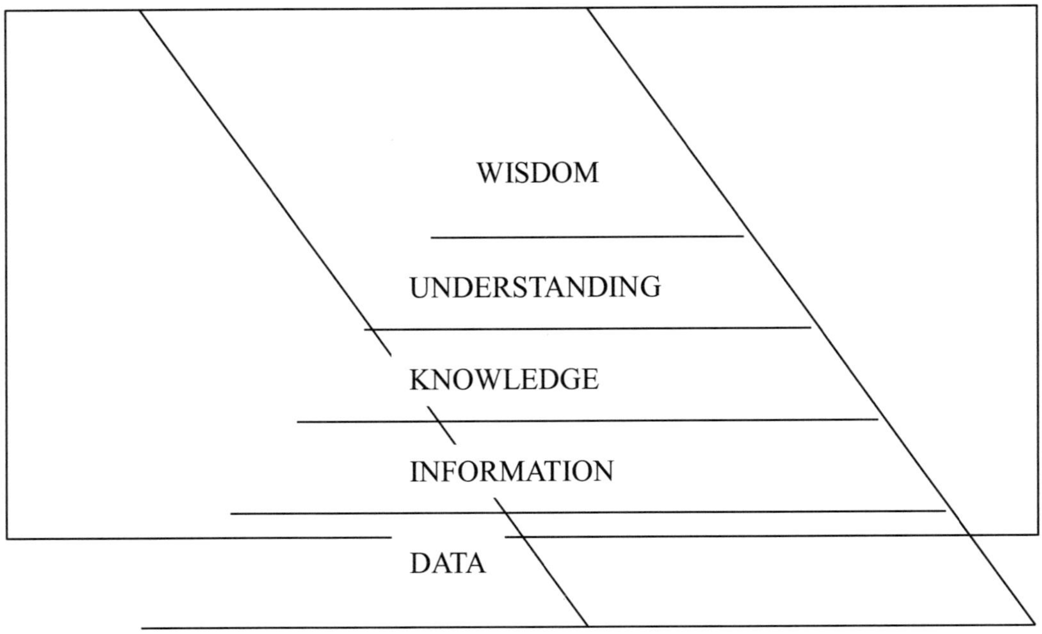

Figure 2. Ackoff's taxonomy summarised.

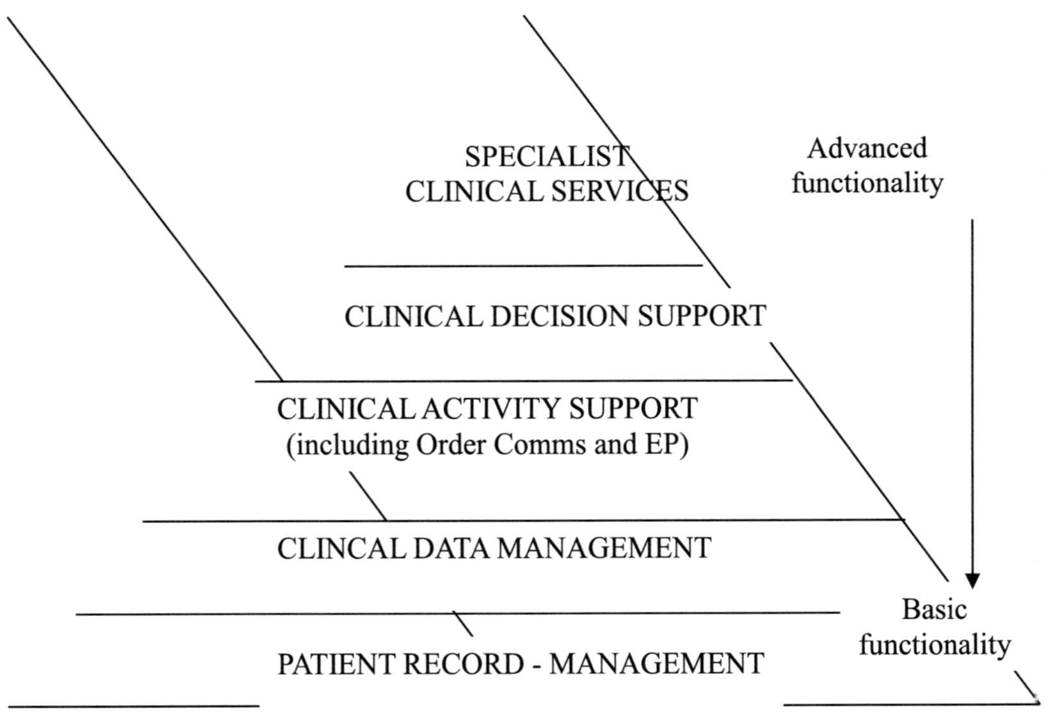

Figure 3. Showing Goundrey-Smith's taxonomy summarised.

http://dx.doi.org/10.5772/intechopen.69769

Activity	Consolidated into group	Sorted into groups
Procurement	Procurement-distribution	Clinical audit
Distribution	Procurement-distribution	Clinical
Dispensing	Dispensing-patient supply	Clinical
Ward stock supply	Procurement-distribution	Clinical
Out-patient services	Dispensing-patient supply	Clinical
Aseptic manufacturing	Dispensing-patient supply	Dispensing-patient supply
Non-aseptic manufacturing	Dispensing-patient supply	Dispensing-patient supply
Medicines information	Clinical	Dispensing-patient supply
Clinical audit	Clinical audit	Dispensing-patient supply
Clinical services	Clinical	Dispensing-patient supply
Pharmacist led clinics	Clinical	Dispensing-patient supply
Record keeping	Management	Dispensing-patient supply
Staff induction	Management	Management
Staff training	Management	Management
Clinical trials	Dispensing-patient supply	Management
Research activity	Clinical	Management
Management of pharmacy	Management	Management
Weekend services	Dispensing-patient supply	Management
On-call services	Dispensing-patient supply	Management
Multidisciplinary team meetings	Management	Procurement-distribution
Finance work	Management	Procurement-distribution
Reports	Management	Procurement-distribution

Table 4. Pharmacy activities consolidated and grouped into types of activity then arranged grouped together.

Most hospitals in the UK will undertake all of these activities to a greater or lesser degree, and the relevant proportions will depend on local circumstances range of specialist services, number of beds in acute hospital etc. The proportion of time relative to their activities by staff might provide a useful indicator. The hierarchy is based on the skill levels of NHS Pharmacy staff routinely undertaking these functions, and this could be reflected in the pay band, which is heavily weighted to qualifications.

1. Specialist clinical service
2. Clinical service
3. Management

4. Dispensing-patient supply
5. Procurement-distribution

Do these functions exhibit any similarities to Ackoff's or Bloom's model? The taxonomies are summarised in **Table 5**.

Superficially, there seems to be similarities with Bloom's than Ackoff's model. However, the real insight comes when the normal amount of time spent on these activities is considered, which can be used as a comparison tool. This starts to lay the basis of assessing the level of 'higher' professional activity.

This method of assessment might be useful for comparing different hospital pharmacies, but the real value of it is in telling an individual pharmacy manager where they might be in terms of how much 'higher' professional activity is going on within a department and what impact implementing technology might have on higher professional functions. Why is this important? Because some surveys indicate that pharmacists feel empowered and have better ward relationship within an EP-RD model. It allows them to 'cover more ground' and enhances professional relationships. Using EP and robotic dispensing potentially reduces the level of dispensing done by pharmacists and therefore moves more towards the apex of the professional pyramid. It does this by reducing the amount of time spent by pharmacy staff on moving medicines about the hospital, thereby releasing staff for other higher skilled work.

These are all strong benefits, both to the pharmacy and patient care and therefore to the organisation. Using the descriptors of the higher elements of Ackoff and Bloom, technology enhances the 'wisdom' and 'creating' levels, increasing the value the pharmacy adds to the organisation. It is in the process of adding extra value to the work done within the pharmacy within resources that makes the service more valuable to the organisation.

The literature raises a further issue. Nation et al. [18] make the point that in the literature there are no common terms used for describing medication errors, and that this makes comparisons in the literature difficult. Goundrey-Smith [10] notes the differences in terminology of EP in the UK and 'computerised physician order entry' (CPOE) in the USA. He points out that the terms are often used synonymously, but that CPOE is a broader term which includes EP, but encompasses transmission of other clinical order types, such as pathology tests or radiology tests. Across Europe, there is a standard that looks at EP in health informatics (European

Suggested hierarchy	Ackoff	Bloom	G'rey-Sth
1. Specialist clinical service	Wisdom	Creating	Specialist clinical
2. Clinical service	Understanding	Evaluating	Clinical decision/support
3. Management	Knowledge	Analysis	Clinical activity support
4. Dispensing-patient supply	Information	Applying	Clinical data management
5. Procurement-distribution	Data	Understanding	Data input/management

Table 5. Summaries of different taxonomies.

http://dx.doi.org/10.5772/intechopen.69769

Committee for Standardisation, European pre-standard (ENV Health informatics); messages for the exchange of information on medicine prescriptions). However, these standards are more to do with message design. There still remains the problem of describing levels of technological functionality across publications. Taxonomies are a means to clarify things or concepts and have an underlying principle for the classification. Castillo et al. [19] did an extensive literature search to look at factors that made physicians use an electronic health record. They identified from their literature search six factors: user attitude towards electronic systems, workflow impact, inter-operability, technical support, communication amongst user groups and expert support. These were defined as the critical adoption factors. However, 55 of the 70 papers reviewed were from the USA, putting a contextual bias to the conclusions. Some UK experiences do not reflect the Castillo et al. [19] view with regards to their six main factors; nor should it. Bell et al. [20] identified 9 EP capabilities. Again, it is a USA-based paper and reflects their context. They proposed five activities in the prescribing pathway that needs to be included in a system. These are as follows:

- prescribe
- transmit
- dispense
- administer
- monitor

From this, they proposed that an out-patient EP system should have the following capabilities: patient selection, diagnosis selection, medication regime selection, safety alerts, formulary alerts, computer assisted dose calculations, transmission to pharmacy (pharmacies) administration and monitoring.

When assessing the literature, there is a problem with regards to evaluating reports of electronic prescribing and dispensing robots. The study by Cantrill et al. [12] looked at EP in three different hospitals in the North-West and assessed the doctors' and pharmacists' views of the benefits of EP. The surveyed hospitals had three different systems. Functionality in the systems was described, apart from the listing in the appendices, but no comments were made regarding the impact of the EP functionality on the results. In reading papers on EP and RD generally, some implicit knowledge of the subject is required because of the lack of common terms for his subject matter. A reader of the Cantrill et al. [12] study unfamiliar with the subject might be misled to regard there being little benefit in developing EP systems in hospitals because it does not consider the impact of different levels of functionality that framed the feedback from pharmacists and doctors.

Another consideration is that as pharmacists become more ward based, they potentially loose dispensary skills. Within EP-RD systems, it is more practical to have staff trained to manage and maintain these systems, but they are not necessarily pharmacists. Technicians can be managing these technical activities. Technicians were taught traditional dispensing skills, but these skills are used much less nowadays. However, as pharmacists move out to the wards, dispensary management, robot management, procurement, IV fluid management and EP dictionary maintenance can be technician led activities. Technology adoption not only increases the potential for increased pharmacist professionalisation but also pharmacy technicians. Technology potentially expands roles (altered boundaries as Barrett et al. put it) [21] but in a positive way. The effect trickles down to non-technical staff. It is possible to create in the UK NHS staff grade structures a clear qualification-based access scheme to allow band 2 staff (unqualified) to become qualified to become band 3 (dispensers), thence to band 4 (basic technician), to band 5 (medicines management technician) and thence to supervisor roles (band 6). This system can be used in part as succession planning to fill vacancies as they arise. It is also cost effective, as replacing a band 5 staff member may only cost after staff have shuffled up a grade a band 2 replacement, as far as the staff budget is concerned. This is a useful rationale when negotiating vacancy replacement with higher hospital management.

The introduction of technology into (health care) professional settings needs to be taken seriously, especially from the perceptions of those who interact with technology as it has the potential to regulate, mediate, govern and represent health care professionals, altering both what they do and who they are. These alterations are neither unidirectional nor deterministic. Health technology opens up a field of multiple possibilities for both re-professionalisation and de-professionalisation. De-professionalisation (e.g. as mentioned above, de-skilling the dispensing process) is conditioned upon four main IT enabled possibilities. First, through automation, technology alters the nature of professional work by shifting its temporal, spatial and manual aspects. Technology through electronic connection has the potential to expand professional boundaries to other occupational groups working at ward level. Pharmacy is no longer just a building on the hospital site that stores medicines, but a comprehensive service every bit as visible as nursing or medical care. Technology enables re-professionalisation in three ways as follows:

- Through automation it undertakes mundane tasks and gives health care professionals the opportunity to undertake more challenging mental and clinical activities in their everyday work.
- By digitalising, transferring and translating information, it expands pharmacists' jurisdictions (through the creation of new or the reshuffling of old responsibilities) and gives them the chance to exercise more discretion and professional judgement.
- Finally, through electronic connection and identification, technology expands professional boundaries by allowing professionals to become a more integrated part of the ward teams.

While there is some agreement about what constitutes professional behaviour, there are no published concepts of what is the professional model in pharmacy or what constitutes a 'higher' professional model. In Canada, Motulsky et al. [22, 23] looked at the impact of technology on professionalisation of community pharmacy and concluded that the adoption of technology increased professionalisation. They defined increased professionalisation as pharmacy practice centred on clinical activities within pharmacy services. The literature abounds with papers written on EP and also on robotic dispensing, but there are few instances where the impact of both together is described. The study by Reifsteck et al. [13] in the USA is a notable exception. There are fewer papers exploring the impact of technology on professionalism

http://dx.doi.org/10.5772/intechopen.69769

within pharmacy. There is also some confusion in the literature regarding what constitutes electronic prescribing, which distorts benefit analysis. This identifies an emerging need for a common language to describe the technology to be able to better compare studies in the literature, akin to Nation et al. [18] call for a common language in medication errors.

Much debate has taken place over what constitutes a profession, a professional, and the concept of professionalism [24–32]). Sociological theorists in the 1960s and 1970s challenged occupations with professional status and the monopoly they hold over society [33, 34]. Denzin and Mettlin in 1968 accused pharmacy of being an incomplete profession based on criteria established by the Trait theory. However, the approach of the sociological theorists could itself be challenged, as the law is quite specific what constitutes a profession: a profession has a registering body, has a code of practice and enforcement arm of the registering body. Breaking the codes of practice can result in being removed from the professional list and unable to practice. This pragmatic assessment seems to have been overlooked by the sociological theorists. However, Pharmacy was labelled (dubiously??) as a quasi profession, in as much it had some but not all characteristics of a profession [27].

Technology opens up a number of, often contradictory, possibilities for shaping pharmacy professionalism but does not determine it. Professionalism (and hence professional satisfaction) is conditioned upon the way in which professionals will exercise their power. The consequences technology imposes on professionals has important implications for how pharmacists use (or not), adapt and adopt technology and points towards an exploratory perspective that looks into how and in what ways technology may shape health care professionals in the future. This has always been a historical tradition, as Anderson's history of pharmacy points out [1].

It becomes apparent that there are implications regarding using the technology and the possible impact on professionalism. The historical summary taken from Anderson's 'History of Pharmacy' frames aspects of professionalism noting that since from 2000 BC, Pharmacy has always been separate from Medicine and is recorded as such. The American Board of Internal Medicine 'Project Professionalism' [35] document and also a similar UK document published in England in 2004 by Rosen and Dewar [36] were both designed to address the teaching of professionalism to undergraduates, against a background that public confidence in the profession needs to be maintained. Similarly for pharmacy, the American Association of Colleges of Pharmacy Council produced a document in 2000 [37] to ensure pharmacy students were taught professionalism in schools of pharmacy, and in the UK, Schafheutle, Hassell Ashcroft et al. produced a similar document in 2010 [38]. These papers were responding to a perceived need, but the earlier publications of these papers in the USA suggest the health care systems in which they operate triggered the need earlier. This might in part be linked to a health care system not free at the point of delivery, but insurance driven, and the need to demonstrate the professions was using expert skill for patient benefit and not exploiting a monopoly situation for their own rewards.

The problem for Pharmacy is that EP and RD developments happen outside the Universities where undergraduates learn their skills, and the opportunities technology creates provides a wide range of opportunities for the profession to develop. The potential to significant increase the efficiency of the dispensing process provides a springboard for increased ward activity. The use of EP to 'police the formulary' removes a small conflict task done by pharmacists, permitting a more positive working relationship with other ward professionals. The potential for reducing the cost of extending the working week in terms of staff numbers also increases availability of pharmacy staff and creates further roles and professional identities for staff. Future digital developments create potentially new professional identities and further develop professional boundaries. The development of an EP-RD system is not only an end in itself, it is potentially a gateway into a more comprehensive and diverse future for all pharmacy staff.

Author details

Roderick J. Beard

Address all correspondence to: rod.beard@chsft.nhs.uk

Pharmacy Department, Sunderland Royal Hospital, Sunderland, UK

References

[1] Anderson S. Making Medicines: A Brief History of Pharmacy and Pharmaceutics. London: Pharmaceutical Press; 2005

[2] Smith JM. Building a Safer NHS for Patients: Improving Medication Safety [Internet]. 2005. Available from: www.doh.gov.uk

[3] Donaldson L. An Organisation with a Memory [Internet]. 2001. Available from: www.doh.gov.uk

[4] Dean BS, Allan EL, Barber ND, Barker KN. Comparison of medication errors between American and UK hospitals. American Journal of Health System Pharmacists. 1995;**52**(22):2543-2549

[5] Karsh BT. Beyond usability: Designing effective technology implementation systems to promote patient safety. Quality and Safety in Healthcare Journal. 2004;**13**:388-394

[6] Cardiff and Vale NHS Trust. Dispensing Error Analysis Scheme (DEAS). Referenced in Building a Safer NHS [Internet]. 1994. Available from: http: www.doh.gov.uk

[7] Swanson D. Automated dispensing: an overview of the types of system available. Hospital Pharmacist. 2004;**11**(2):66-68

[8] Whittlesea C, Phillips C. Automated dispensing—how to evaluate its impact. Hospital Pharmacist. 2004;**11**:283-285

[9] James L. Role Conflict: Occupational Stressors vs Patient Safety [Internet]. 2010. Available from: www.kcl.ac.uk

[10] Goundrey-Smith S. Principles of Electronic Prescribing. London: Springer-Verlag; 2008

[11] Goundrey-Smith S. Electronic prescribing. Pharmaceutical Journal. 2008;**280**:599-602

[12] Cantrill JA, Abdel-Qadar DA, Tully MP. Satisfaction predictors and attitudes towards electronic prescribing systems in three UK Hospitals. Pharmacy World & Science. 2010;**32**:581-593. http://www.springerlink.com/content/k0p6xn36124w2j30/

[13] Reifsteck M, Swanson T, Dallas M. Driving out errors through tight integration between software and automation. Journal of Healthcare Information Management. 2006;**20**:35-39

[14] Gonidec P, et al. Performance of an automated dispensing system combined with a computerised prescription order entry. Annales Pharmaceutiques Francaises. 2009;**67**:84-90

[15] Beard J. Wood D. Application of lean principles can reduce inpatient prescription dispensing times. Pharmaceutical Journal. 2010;**284**:369-371

[16] Bloom B. Taxonomy of Educational Objectives: The Clarification of Educational Goals. London: Addison Wesley; 1956

[17] Ackoff RL. From data to wisdom. Journal of Applied Systems Analysis. 1989;**15**:3-9

[18] Nation RL, et al. Multiplicity of medication safety terms, definitions and functional meanings: When is enough, enough? Quality Safe Health Care Journal. 2005;**14**:358-363

[19] Castillo VH, et al. A knowledge-based taxonomy of critical factors for adopting electronic health record systems by physicians. A systematic literature review. BMC Medical Informatics and Decision Making. 2010;**10**:50

[20] Bell DS, et al. A conceptual framework for evaluating out-patient electronic prescribing systems based on their functional capabilities. Journal American Medical Informatics Association. 2004;**11**:60-70

[21] Barrett M, et al. Reconfiguring boundary relations: Robotic innovations in pharmacy work organisation. Science. 2012;**23**(5):1448-1466

[22] Motulsky A, et al. The impact of electronic prescribing on the professonalisation of community pharmacists: A qualitative study of pharmacists' perception. Journal of Pharmacy and Pharmaceutical. 2008;**11**(1):131-146

[23] Motulsky A, et al. Electronic prescriptions and disruptions to the jurisdiction of community pharmacists. Social Science & Medicine. 2011;**73**(1):121-128

[24] Hammer DP. Development and testing of an instrument to assess behavioural professionalism of pharmacy students. American Journal of Pharmaceutical Education. 2000;**64**:141-151

[25] Hammer DP. Professional attitudes and behaviours: The "As and Bs" of professionalism. American Journal of Pharmaceutical Education. 2000;**64**(4):455-464

[26] Hammer DP, et al. Student professionalism. American Journal of Pharmaceutical Education. 2003;**67**(1/4):544-573

[27] Joyce MC. What does professionalism really mean? Journal of the American Pharmaceutical Association. 2002;**42**(5):677-678

[28] Pharmacy and the concept of professionalism. Journal of American Pharmaceutical Association. 1968;**8**(issue 5):228-230

[29] Pierpaoili MS. What is a professional? Hospital Pharmacy Times May. 1992. pp. 9-10

[30] O'Toole PT, Kathurian N, Mishra M, Schuket D. Teaching professionalism within a community context: Perspectives from a national demonstration project. Academic Medicine. Journal of the Association of American Medical Colleges. 2005;**80**(4):339-343

[31] Hilton S. Medical professionalism: How can we encourage it in our students? The Clinical Teacher. 2004;**1**(2):69-73

[32] Tordoff A, et al. Definitions of professionalism in pharmacy practice: A systematic review. International Journal of Pharmaceutical Practice. 2009;**17**(S1A):35-36

[33] Wagner P, et al. Defining medical professionalism: A qualitative study. Journal of Medical Education. 2007;**41**(3):288-294

[34] Denzin NK, Metlin CJ. Incomplete professionalisation: The case of pharmacy. Social Forces. 1968;**46**:375-381

[35] American Pharmaceutical Association. APhA-ASP/AACP-COD Task Force on professionalism. White paper on pharmacy student professionalism. Journal of the American Pharmaceutical Association. 2000;**40**:96-100

[36] Rosen R, Dewar S. On Being a Doctor: Redefining Medical Professionalism for Better Patient Care. London: The King's Fund; 2004

[37] APhA-ASP/AACP Committee on Student Professionalism. Pharmacy Professionalism Toolkit for Students and Faculty [Internet]. 2004. Available from: http://www.pharmacist.com/AM/Template.cfm?Section=Professionalism_Toolkit_for_Students_and_Faculty [Accessed: 20 July 2009]

[38] Schafheutle E, et al. Professionalism in Pharmacy Education. London: Pharmacy Practice Research Trust; 2010

Section 6

Epilogue

Chapter 9

Human, Not Humanoid, Robots

Domenico Parisi

Additional information is available at the end of the chapter

http://dx.doi.org/10.5772/intechopen.70117

Abstract

Robots that resemble human beings can be useful artefacts (humanoid robots) or they can be a new way of expressing scientific theories about human beings and human societies (human robots), and while humanoid robots must necessarily be physically realized, human robots may be just simulated in a computer. If the simulated robots do everything that human beings do, the theory which has been used to construct the robots explains human behaviour and human societies. This chapter is dedicated to human robots and it describes a number of individual and social human phenomena that have already been replicated by constructing simulated human robots and simulated robotic societies. At the end of the chapter, we briefly discuss some of the problems that human robots will pose to human beings.

Keywords: human robots, humanoid robots, science, technology

1. Human robots as a new science of human beings

"Robot" is an ambiguous word. It has two different meanings. Robots can be physical artefacts with practical applications and economic value or they can be a new science of human beings. For robots as practical artefacts, success is that there are people who are disposed to spend their money to buy them. For robots as science, success is to construct robots that do *everything* that human beings do, because only if scientists are able to construct robots that do *everything* that human beings do, they will finally understand human beings. Since practically useful artefacts that look like human beings and do some of the things that human beings do are called "humanoid" robots, to make the distinction explicit, we will call robots as science "human" robots. In this chapter, we will be concerned with human, not humanoid, robots.

While humanoid robots necessarily are physical robots, human robots may be just simulated in a computer. In fact, human robots are based on the general assumption that the best way for science to understand X is to simulate X in a computer. If, when the simulation runs in the computer, its results correspond to what scientists empirically know about X, they are entitled to conclude that the theory incorporated in the computer program captures the mechanisms and processes which underlie X and, therefore, it explains X. Computers can be useful to science in many other ways but they are a true scientific revolution if they are used to express scientific theories in a novel way. So far, scientific theories have been expressed using mathematical symbols or using words. Physicists express their theories using mathematical symbols. Scientists who study human beings express their theories by using words—the only exception is economics but economics is not a science but an applied discipline—and words are a problem for science because they tend to have unclear meanings and to mean different things to different scientists—and defining one word by using other words clearly does not solve the problem. The consequence of expressing scientific theories by using words is that scientists rarely agree on the empirical predictions that can be derived from a theory and they spend most of their time to do endless discussions which resemble more those of philosophers than those of scientists. Human beings are more complicated and more difficult to study than nature, but it is the fact that scientists who study human beings express their theories by using words, which is the real reason why the sciences that study human beings and human societies are so much less advanced than the sciences that study nature.

Computer simulations solve the problem. A theory is formulated as a computer program and, when the program runs in the computer, it generates a large number of quantitative results, which are the predictions derived from the theory/simulation. If these results correspond to what scientists empirically know about reality, the theory/simulation is confirmed. If they do not correspond, the theory/simulation must be modified or abandoned.

Human robots are computational theories of human beings. To understand human beings, scientists must construct simulated human beings which are like real human beings and which do all that real human beings do. Humanoid robots reproduce only an extremely limited number of things that human beings do. They have a body which has some external resemblance to the human body, they walk on two legs, they reach and grasp objects with their hands, they express emotions with their face which they do not really feel, and they produce words which they do not really understand. Human robots must have a body which does not only have a human-like external form but also contains internal organs and systems that simulate the internal organs and systems of the human body—not only the brain, but also the heart, the lungs, and the visceral, endocrine, and immune systems [31]. They must be the result of a process of evolution that takes place in a succession of generations and of a process of development and learning that takes place during the course of a robot's life [1, 2]. Each individual robot must be different from all other robots and the robots must have all sorts of pathologies both of the body and of the mind. They must have their own independent motivations and their behaviour must be determined by their motivations. They must actually feel the emotions which they express with their face, their voice, and their body. They must talk to other robots by producing sounds that they actually understand, and they must also talk to themselves without producing audible sounds (think) [25]. They must respond to stimuli which do not arrive to the brain from the external environment or from their own body [27]

http://dx.doi.org/10.5772/intechopen.70117

but are self-generated by their own brain (mental life). They must be born from a female and a male robot, they must live for a certain period of time, and then they must die. And they must be very social. They must live in families, they must have friends, they must cooperate or compete with other robots, and they must organize themselves in societies that have economic and political institutions and that change historically. They must learn by imitating other robots and develop cultures that may change from one generation to the next. They must modify the environment in which they live, they must make artistic artefacts and must expose themselves to the artistic artefacts made by other robots, and they must have religion, philosophy, and science.

But human robots are not only theories. Scientists can also do experiments on their simulated human beings. They can vary the value of the different variables and see the consequences of these variations. Laboratory experiments are a very important scientific tool but, while they are a perfect tool in the hands of physicists, chemists, and biologists, they have many limitations when they are used to study human beings and human societies. Most of what psychologists know about human beings is derived from laboratory experiments but laboratory experiments provide them with a very limited knowledge of human behaviour [3]. First, human behaviour is the result of the interactions of human beings with their environment, but the laboratory is a very simplified environment, which is very different from the real environment. Therefore, what human beings do in an experimental laboratory may be very different from what they do in their real life. Second, outside the experimental laboratory human beings do what they want to do, whereas experimental subjects do what the experimenter wants them to do.

The problem is even more serious for social scientists—anthropologists, sociologists, economists, and political scientists. Social scientists do very few laboratory experiments because the social environment of human beings is almost impossible to reproduce in the laboratory and because social phenomena are very complex and they are much more extended in time than laboratory experiments. Therefore, social scientists are mostly limited to collecting statistical data on the consequences of human behaviour, making interviews, and reading the books of other social scientists.

A robotic science of human beings changes all of this. Since the behaviour of human beings depends on the environment in which they live what observable and measurable aspects of the robots', to understand human behaviour scientists must simulate in the computer not only human beings but also their natural and social environment [4]. And they not only simulated *laboratory* experiments but also simulated *ecological* experiments in which they vary the natural and social environment that their human robots live and see how the robots' behaviour depends on the particular environment in which they live. And they can also do counterfactual experiments. They can let the robots live in an environment which does not exist and see whether the robots behave in the non-existing environment as predicted by their theories.

Expressing scientific theories as computer simulations has another important advantage. Science is divided into disciplines, with some disciplines studying some of the phenomena that make up reality and other disciplines studying other phenomena. The problem is that reality is not made up of separate classes of phenomena but it is a large ensemble of phenomena which are all connected together and, often, to understand the phenomena studied by one discipline it is necessary to take into account the phenomena studied by other disciplines. Today, there are attempts at addressing this problem by doing what is called *inter-disciplinary* research: scientists of different disciplines discuss and collaborate together to better understand the phenomena which interest them. But inter-disciplinary research does not really solve the problem because it continues to take scientific disciplines as given, with each discipline studying one particular class of phenomena and possessing theories that try to explain that particular class of phenomena.

Both the science of nature and the science of human beings are divided into disciplines but for the science of nature the division into disciplines is not really a problem because physics, chemistry, and biology use the same empirical methods, have very similar conceptual and theoretical traditions, and share a view of nature as made up of physical causes that produce physical effects and as possessing an inherently quantitative character. On the contrary, the division of the science of human beings into disciplines—psychology, anthropology, linguistics, sociology, economics, political science—has very negative consequences because these disciplines do not share the same empirical methods, have very different conceptual and theoretical traditions, and do not have a unified view of the phenomena they study.

Computers change this situation because they make it possible to develop a non-disciplinary science of reality, a science which completely abolishes scientific disciplines. Science is divided into disciplines because scientists are human beings and their brain is too small to formulate theories that take into account and try to explain the data collected by different scientific disciplines. Computers have a much larger and more powerful "brain," with a memory that can contain enormous quantities of data and a computing capacity that can take into account all the relations among the data. For practical reasons, empirical data will continue to be collected by different scientists but the theories-simulations that explain these data and make predictions about them will not be physical, chemical, biological, psychological, or social theories but they will simply be theories of reality.

A robotic science of human beings is a non-disciplinary science that will abolish not only the divisions among the disciplines that study human beings and their societies but also the great division between the disciplines that study nature and those that study human beings and their societies—which is the most serious obstacle to a scientific comprehension of human beings. Clearly, the creation of a non-disciplinary science of human beings will be a gradual process. A robotic science of human beings will begin by constructing robots and societies of robots, which greatly simplify with respect to real human beings and real human societies, but then the robots and the robotic societies will become progressively more complex and more similar to real human beings and real human societies. Today, only some psychologists and some neuroscientists are interested in human robots but human robots will progressively interest, on one side, biologists and chemists and, on the other side, anthropologists, sociologists, economists, political scientists, and even historians.

To realize a complete robotic science of human beings, it is possible to adopt two different strategies. One strategy is based on the principle "one robot/one phenomenon" and adopting

http://dx.doi.org/10.5772/intechopen.70117

this strategy means to construct different robots each of which reproduces in a more realistic way one single aspect of human behaviour. The other strategy is based on the principle "one robot/many phenomena," and adopting this strategy means to construct one and the same robot that reproduces in a more simplified way many different human phenomena and then to progressively add more details and make the robot more realistic. The second strategy is better than the first one because one and the same human being perceives what is in his or her environment, moves his or her body, remembers, predicts, speaks and understands, thinks, has a variety of motivations and emotions, does things with other human beings, and participate in the creation and functioning of social structures. Therefore, one and the same robot must do all these things.

If this is the final goal of a robotic science of human beings, this science poses a very general and interesting question. Human robots are theories that try to explain human beings by simulating them in a computer, and they are one example of a general principle, which I think in the future will be adopted by all scientists, according to which, whatever phenomenon science wants to explain, what science must do is simulate the phenomenon in the computer. But there is an important difference between scientific theories expressed by using words or mathematical symbols and theories expressed as computer programs. Verbal and mathematical theories *necessarily* simplify with respect to the phenomena they want to understand and their value depends on the goodness of these simplifications. Theories expressed as computer programs begin by reproducing reality in a very simplified way but then scientists can add more and more details until the simulation *completely* replicates reality. What are the consequences of this progressive convergence between scientific theories and reality? When should scientists stop adding more details? I don't know what is the answer to this question, and I wait for suggestions from philosophers of science.

2. Human robots are a non-verbal science of human beings

Human robots pose this and other interesting philosophical problems but understanding human beings by constructing human robots is the opposite of doing philosophy. While philosophy is made of words and of discussion about words, robotics has no use for words. Psychologists and social scientists use words to formulate their theories, and many of these words have a philosophical origin or have been discussed for centuries by philosophers: sensation, perception, attention, memory, thinking, reasoning, planning, motivation, emotion, representation, concept, category, meaning, object, property, action, intention, goal, consciousness, norms, and values. Robotic scientists can use these words only if they can point out what observable and measurable aspects of the robots' behaviour, brain or society they call sensation, perception, attention, memory, motivation, emotion, etc.

Take the word "category," an important word for both psychologists and philosophers. A robot can be said to have categories if it behaves in the *same* way towards *different* things. Here is a very simple example [5]. A population of robots lives in an environment which contains both roundish and angular objects but no two objects have exactly the same shape. The roundish objects are food and the angular objects are poison and, to remain alive and have offspring, the robot must reach and eat the roundish food objects and avoid the angular poison objects. If, when we look at the robots on the computer screen, we see that the robots approach and eat the roundish objects and avoid the angular objects, we are entitled to say that they possess the category of food and the category of poison because the word "category" is defined not by using other words but by looking at the robots' behaviour.

And the word "category" can be defined not only by looking at the robots' behaviour but also by examining the robots' brain. The robots' brain is a neural network made of artificial neurons with a level of activation that varies from one cycle of the simulation to the next cycle and of connections between neurons through which one neuron influences the level of activation of another neuron. Each connection has a quantitative weight which can be either a positive number (excitatory connection) or a negative number (inhibitory connection), and it is this weight that determines how the activation level of one neuron influences the activation level of another neuron. The brain of our robots is made of three types of neurons—visual neurons, internal neurons, and motor neurons—and since the visual neurons are connected to the internal neurons and the internal neurons are connected to the motor neurons, what a robot sees determines what the robot does. If we call "pattern of activation" the ensemble of levels of activation of a set of neurons in each cycle the pattern of activation of the visual neurons is caused by the shape of the object that the robot is currently seeing, this pattern of activation causes a pattern of activation in the internal neurons which, in turn, causes a pattern of activation in the motor neurons, and the pattern of activation of the motor neurons causes the robot to approach or avoid the object.

At the beginning of the simulation, the connections of the robots' neural network have random weights and, therefore, the robots are unable to distinguish between the roundish and the angular objects and to approach the roundish objects and avoid the angular objects. Therefore, on average, these robots do not eat much food and they also eat some poison, which means that they have a short life and are unable to generate many offspring.

The capacity to distinguish between the roundish and the angular objects is acquired through a process that takes place in a succession of generations and simulates biological evolution. The selective reproduction of the robots which, for purely random reasons, have better connection weights in their neural network and, therefore, have some tendency to approach the roundish objects and to avoid the angular objects, and the addition of random changes in the quantitative weights of the connections inherited by the offspring robots from their parent robots (genetic mutations)—which in some cases can result in offspring which are better than their parents—determine, in a succession of generations, the progressive acquisition of the capacity to approach and eat the roundish objects and to avoid the angular objects. Therefore, at the end of the simulation, we can say that the robots have acquired the category of food and the category of poison.

This is what we find when we examine the robots' behaviour. But we can also ask: What happens in the robots' brain that make the robots approach and eat the roundish objects and avoid the angular objects? To answer this question, we look at how the different objects are "represented" in the robots' brain, where the neural "representation" of an object is the pattern of activation of the internal neurons of a robot's neural network which is caused by the sight of the object. What we find is that while in the robots of the first generation the neural representations of the roundish and angular objects are confused together; after a certain number of generations, the roundish objects cause very similar patterns of activation in the internal neurons and the same for the angular objects, but the patterns of activation caused by the roundish objects are different from the patterns of activation caused by the angular objects. This means that the robots have evolved the capacity to categorize some objects as roundish and other objects as angular.

We have described this simulation to illustrate how a robotic science of human beings treats words. Robotic scientists can use words—in our case, the word "category" and the word "representation"—but only if they can point out to what these words refer to either in the robots' behaviour or in the robots' brain. As we have already said, this is not what happens in the traditional sciences that study human behaviour and human societies. Scientists dedicate much of their time to defining words by using other words and to discussing the meaning of a word without never reaching an agreement. The consequence is that from a verbally formulated theory different scientists may derive different predictions and, therefore, their theories can never be confirmed or disconfirmed by what is empirically observed and measured. By not using words or by using words only if their meaning can be translated in what is observed and quantitatively measured, the robotic science of human beings solves this problem.

3. Only a robotic science of human beings can look at human beings with the detachment required by science

Scientists are human beings and, unlike when they study nature, when they study human beings they are almost inevitably influenced by their values, desires, and fears. Therefore, from a verbally formulated theory, scientists may not only derive different empirical predictions because the theory is unclear and ambiguous but they may also be influenced by their values in choosing which predictions to derive from the theory—which is another reason why the sciences that study human beings and human societies are so much less advanced than the sciences that study nature.

This changes if scientists express their theories of human beings and human societies by constructing human robots and human robotic societies. What the robots do and why they do are under the eyes of everyone and scientists cannot deny the evidence provided by the robots. This is another important advantage, which is provided by a robotic science of human beings and human societies. This science will make it possible to study human beings and human societies with the same detachment with which natural scientists study nature.

A related problem is that scientists belong to different cultures and, while this has no consequences when they study nature and when they study human beings and human societies, they tend to be influenced by their culture. This is very clear for anthropologists but it is a general problem for the sciences that study human beings and human societies because science must be universal and independent from culture. Studying human beings and human societies by simulating them in a computer solves this problem. By constructing robotic societies that have different cultures, scientists will be able to look at human beings and their cultures—including their own culture—with the necessary detachment.

4. Human robots must have their own motivations and they must do they want to do

Although human robots will make it possible for science to know human beings much better than its previous attempts at knowing them, they will also pose many problems to human beings. Robots as technologies already pose problems to human beings but, since these problems are discussed in the other chapters of this book, we will concentrate on the problems that robots as science will pose to human beings.

The most serious of these problems is due to the fact that while humanoid robots are constructed to satisfy *our* motivations, human robots must have *their* own motivations and they must do want *they* want to do, not what *we* want them to do. Some humanoid robots are said to be "autonomous" but, since humanoid robots are technological artefacts, technological artefacts cannot be really autonomous. They can autonomously decide what to do to reach a certain goal but the goal is decided by us. A humanoid robot can autonomously decide how to move its arm and its fingers to reach and grasp an object with its hand but *we* decide that it must reach and grasp the object with its hand. Therefore, humanoid robots can be cognitively (behaviourally) but not motivationally autonomous. Human robots must be both cognitively and motivationally autonomous because human beings are both cognitively and motivationally autonomous. They must decide both that they want to reach and grasp the object with their hand and know how to move their arm and their fingers to reach and grasp the object.

Motivations are the most important component of human behaviour—and of the behaviour of all animals. One often hears that behaviour is caused by stimuli, but this is not true. An individual's behaviour is *guided* by stimuli but it is *caused* by the individual's motivations. The robots described in Section 2 had the motivation to eat food and the motivation not to eat poison, and the real cause of their behaviour was these two motivations. Seeing a roundish object or an angular object only guided them towards the roundish object or away from the angular object.

Since the motivations of those robots were only two and they always had the same strength, it was rather easy for the robots to decide which of the two motivations to satisfy with their behaviour at any given time: seeing a roundish objects activated one motivation and seeing an angular object activated the other motivation. Human beings have a much greater number of different motivations and the strength of these motivations can change from one moment to the next as a function of various factors. Therefore, it is more difficult for human beings to decide which motivation they should try to satisfy with their behaviour at any given time. Their motivations lie dormant in their brain and in their body and they are activated not only by the external stimuli—like the two motivations of the robots of Section 2—but also by stimuli self-generated inside their brain and inside their body. The problem is that human beings—and all animals—cannot satisfy two or more motivations at the same time and, therefore, in any given moment, they must decide which of their different motivations they should try to satisfy with their behaviour. Since their motivations have different strengths and this strength varies with the circumstances, they try to satisfy the motivation which at any given time has the greatest strength.

This is a simple example of robots that have two motivations whose strength varies from time to time [6]. The robots need both energy and water to remain alive and, since their body constantly consumes both energy and water, they must both eat food (green objects) and drink water (white objects). The robots' body contains two internal stores, one for energy and the other one for water, and the robots' brain has two additional sets of sensory neurons whose activation level reflects the quantity of energy and the quantity of water currently contained in the two bodily stores. These neurons are activated when the quantity of energy or water contained in the robots' body is below a certain level and it is their activation that makes the robots feel hungry or thirsty. The capacity of the robots to respond to hunger by looking for food and to thirst by looking for water evolves in a succession of generations. At the beginning of the simulation, the robots do not look for the green objects when they feel hungry and for the white objects when they feel thirsty but, after a certain number of generations, the robots look for food and ignore water when there is little energy in their body and they feel hungry and they look for water and ignore food if there is little water in their body and they feel thirsty.

Although motivations, not external stimuli, are the real causes of behaviour, external stimuli have an important motivational role because they may activate different motivations. For example, the sight of a predator may activate in a robot the motivation to fly away from the predator while the sight of a robot of the opposite sex may activate the motivation to mate with the robot of the opposite sex. This is true for both animal robots and human robots. But human robots must be more complex because their motivations must be activated not only by external stimuli (the sight of a predator robot or the sight of a robot of the opposite sex) or by internal stimuli self-generated by their own body (hunger and thirst) but also by internal stimuli self-generated by their own brain (thoughts, memories, and imaginations).

But human robots must not only have their own motivations. They must also feel emotions [32] because emotions are a submechanism of motivations [7]. Emotions are states/processes of the body/brain that increase the current strength of one particular motivation so that the individual will choose to satisfy this motivation rather than other motivations. Robots which feel emotions are robots whose brain includes a set of neurons that function differently from the other neurons. First, when they are activated, their activation persists for a certain number of input/output cycles and, second, they send stimuli to other organs and systems that are inside the body such as the heart and the visceral system [31] and these other organs and systems respond by sending stimuli to the brain which modify the strength of the various motivations. This emotional circuit makes the motivational choices of the robot more adaptive—although they may also cause psychical disturbances, for example, if the robot finds it impossible to satisfy a motivation which, for the robot, has a very high strength.

Here is one example of how emotions can help robots to take better motivational decisions [8]. The robots live in an environment which not only contains food objects that they must eat to remain alive but also contains a predator that can suddenly appear and kill the robots. For adaptive reasons, the motivation to fly away from the predator is intrinsically stronger than the motivation to eat and, in fact, when the predator appears, the robots cease to look for food and they fly away from the predator. We compare two populations of robots. The neural network of the robots of one population has only sensory neurons for food and sensory neurons for the predator, whereas the neural network of the robots of the other population, in addition to these sensory neurons, has a set of emotional neurons. These emotional neurons are not activated by the sight of food but they are only activated by the sight of the predator, and their activation persists even if the robot flies away and, therefore, it ceases to see the predator. Since these emotional neurons send their connections to the motor neurons, they influence the robots' behaviour.

When we compare the two populations of robots, we find that the robots with the emotional neurons are less likely to be killed by the predator compared to the robots without the emotional neurons. If we look at the robots' behaviour on the computer screen, we see that the robots with the emotional neurons immediately run away from the predator as soon as they see the predator and they continue to run away even if they cease to see the predator, whereas the robots without the emotional neurons are less good at flying away and, therefore, they are more easily killed by the predator. The robots with the emotional neurons in their neural network can be said to experience the emotion of fear, and experiencing the emotion of fear helps them to remain alive.

Here is another example that demonstrates how feeling emotions helps the robots to take better motivational decisions. The robots we have described so far do not have a sex and they do not need a mate to generate offspring. The new robots are males and females, and to generate offspring, a robot must mate with a robot of the other sex. (The male robots look differently from the female robots.) This means that these robots also have two motivations to satisfy, the motivation to eat to remain alive and the motivation to mate to have offspring, and they must divide their time between looking for food and looking for a robot of the opposite sex. Again, we compare a population of robots with a set of emotional neurons in their brain and another population of robots without emotional neurons. The results are that the robots with the emotional neurons in their brain are more attracted by the robots of the opposite sex and, therefore, they have more offspring than the robots without the emotional neurons. They eat what is sufficient to remain alive but, unlike the robots without the emotional neurons, as soon as they see a robot of the opposite sex, they ignore food and approach the robot of the opposite sex. Unlike the robots without the emotional neurons, they can be said to experience the emotion of "sexual attraction."

Like motivations, emotions clearly distinguish between robots as science and robots as technology, between human and humanoid robots. Some of today's humanoid robots *express* emotions with the movements of their face or with the tone of their voice because this makes them more attractive for potential buyers, but they do not really feel these emotions. Theirs are *unfelt* emotions—an obvious contradiction. On the contrary, human robots must actually "feel" emotions because human beings actually feel emotions, and they must express their emotions with their face, voice, and body but also keep their emotions for themselves because this what human beings do.

Robots that have their own motivations and emotions contradict Asimov's three laws of robotics. They must do what they want to do because human beings do what they want to do and they cannot obey laws unless they themselves promulgate these laws because human beings obey (most of the times) laws that they themselves have promulgated. In fact, human robots are not really robots if the word "robot" must continue to have its original meaning of "slave worker," because human beings are not slave workers.

http://dx.doi.org/10.5772/intechopen.70117

5. Human robots must be social robots

Another characteristic of human robots that will pose problems to human beings is that human robots will need to be very social robots because human beings are very social animals. Human beings live with other human beings, they spend most of their life doing things with other human beings, they have cultures that make them behave and think like some other beings but unlike other human beings, and they have economic and political institutions. Therefore, human robots must live and interact with other robots, they must talk with other robots, they must live in societies that are like human societies, and they must develop cultures.

Although today one often hears of social robots, social robots are not really social because they interact with us, not between them—and the reason is obvious. Today's "social" robots are constructed to take care of old or ill human beings, to entertain human beings of all ages, and to do other things with human beings because this is what makes it possible to sell them and produce profits. But they do not interact with other robots. The only robots which interact with other robots are those of swarm robotics but the robots of swarm robotics not only resemble much simpler animals than human beings but the robots that make up a swarm of robots are all identical and for them success is only collective success, while no two human beings—and no two members of the any animal species—are identical and a crucial factor in social life is the contrast between individual and collective success.

In fact, a robotic social science that lets us better understand the enormous variety of human social phenomena still does not exist. Today, some human social phenomena are simulated in the computer by using "agents," not robots. Agents do not have a body, do not have a brain, and they do not live in a physical environment. They receive abstract inputs from other agents and, on the basis of very simple rules, they respond by sending abstract inputs to other agents. Agent-based social simulations are useful tools but they must be seen as only a first step towards a robotic social science. If we want to really understand human social behaviour, we must replace agents with robots because human beings do not cease to have a body and a brain and to live in a physical environment when they interact with other human beings and create societies and cultures [9–12, 23, 28].

In this section, we describe robots that simulate some very basic aspects of human sociality but, since human sociality is very complex, most of the work remains to be done.

A very important aspect of human social behaviour is language. Human beings interact together by using language and, therefore, human robots must have language. Humanoid robots seem to understand the linguistic sounds that they hear and the linguistic sounds that they themselves produce but this is not really true. They are only programmed to respond in specific ways to specific sounds and to produce specific sounds in the appropriate circumstances. To have language is something different. It is to possess a neural network which, in addition to sensory and motor neurons, has two sets of reciprocally connected internal neurons. The patterns of activation of the first set of internal neurons are the neural representations of the different objects which the robot sees, whereas the patterns of activation of the second set of internal neurons are the neural representation of the different sounds which the robot hears. The robot learns language in a succession of trials and, at the end of learning, since the two sets of internal neurons are reciprocally connected, seeing an object causes the appearance of the neural representation not only of the object but also of the sound that designates the object (speaking) and hearing a sound causes the neural representation not only of the sound but also of the object which is designated by the sound (understanding) [18, 19, 26, 30].

What difference does it make to have language? To answer this question, we return to the robots we have described in Section 2. To remain alive and reproduce, those robots had to distinguish between two categories of objects, roundish (food) and angular objects (poison), and to eat the first category of objects and avoid the second category of objects. Now we teach these robots to understand language and we find that if during their life these robots learn to respond to one sound ("food") by approaching and eating the roundish objects they see and to respond to a different sound ("poison") by avoiding the angular objects they see, they live a longer life and have more offspring. Why? If we examine the neural networks of the robots, we find that the neural representation of the roundish object is more similar than they were for the robots without language and the same for the neural representation of the angular objects. Language makes behaviour more effective.

Of course, language has many other uses and many other aspects. We have constructed robots that illustrate some of these other uses and aspects [24, 25] but, again, most of the work is still to be done.

We now turn to other aspects of human sociality and we begin by describing robots which, like human beings, are males and females and, to reproduce, must mate with a robot of the other sex [13]. Male and female robots have different colours and this makes them recognizable as males or females by the other robots. But the real difference between male and female robots is that, after mating with a female robot, a male robot can immediately reproductively mate with another female robot and generate other offspring, whereas female robots have a period during which they are non-reproductive due to pregnancy, hormonal changes, lactation, and other factors and also their colour changes so that males can distinguish them from non-pregnant females. Both male and female robots do not have only the motivation to mate and have offspring but they also have the motivation to eat because if they don't eat, they die. The question is: What motivation is stronger, mating or eating?

The answer depends on the sex of the robots. At the end of the simulation, we bring the robots, one at a time, into an experimental laboratory and we let them choose between two alternatives. The results are the following. If male robots must choose between a piece of food and a reproductive female, almost all male robots prefer the non-reproductive female to the piece of food. Why? The answer is that, while in the robots' environment food is always available, this is not true for reproductive females because at any given time many female robots are non-reproductive. Therefore, unless they are very hungry, male robots are more interested in reproductive females than in food. On the contrary, if male robots must choose between food and a non-reproductive female, they almost completely ignore the non-reproductive female and they choose food.

Female robots do not only behave differently from male robots but they also behave differently when they are reproductive and when they are non-reproductive. If reproductive females must choose between food and a male robot, they tend to choose food rather that the male robot, and this implies a strategy of using one's time to look for food and simply waiting for a male to mate with because males are always looking for non-pregnant females. But what is interesting is that the same happens if a non-pregnant female must choose between a male and another non-pregnant female. The non-pregnant female prefers the non-pregnant female to the male. Why? Perhaps because, in the real environment, staying close to other non-pregnant females makes non-pregnant females more attractive for males. Ignoring males is even more frequent among non-reproductive females. A non-reproductive female must choose between a male and food or between a non-reproductive female and food, almost always chooses food.

The next step is families. Families are groups of genetically related individuals who live together and, since families are a very important human social phenomenon, human robots must live in families. The members of a family—mother, father, daughters, sons, grandmothers, grandfathers—live together because by living together they can help each other, and they are motivated to help each other because this increases the probability that their genes or the copy of their genes possessed by their relatives will remain in the genetic pool of the population (kin-selection).

We have simulated some simple phenomena concerning human families. In one simulation, when they are very young and therefore they are still unable to find the food which exists in the environment, the robots evolve the behaviour to follow their parents rather than other robots because, in parallel, parents have evolved the behaviour of feeding their very young offspring. In another simulation, sisters and brothers evolve the behaviour of giving some of their food to their sisters and brothers but not to extraneous robots and, in a third simulation, grandmothers and grandfathers evolve the behaviour of feeding their nephews even if this may cost them their life.

Other social phenomena go beyond families and concern entire communities. Social proximity is (or was) a pre-condition for social interaction and it may be influenced by the nature of the environment. Consider two environments. In one environment, food exists in all parts of the environment, whereas in the other environment food only exists in certain parts of the environment. What we find is that while the robots of the first environment do not live near to one another, the robots of the second environment live together in communities in those parts of the environment that contains food [14]. But robots may live near to one another independently of the nature of the environment because, if they live near to one another, they may coordinate their behaviour and display useful collective behaviours [15, 16].

Human beings can live in smaller or larger communities and human history is characterized by the progressive increase in the size of human communities to the point that, today, human beings tend to live in a single global community. To reproduce this phenomenon, we compare two populations of robots both living in a seasonal environment. The robots of one population are divided into a certain number of small communities, each living in its small territory, whereas the robots of the other population are a single community and their territory is the entire environment. The results of the simulation are that the robots that form a single large community and go everywhere in the environment looking for food continue to exist, whereas the robots that are divided into small communities become extinct.

The robots I have described so far need only one type of food to remain alive. However, if to remain alive the robots need to eat two different types of food and the two types of food are in two different parts of the environment, the robots must continuously move from one to the other part of the environment, and this is very expensive in terms of both time and energy. In these circumstances, the robots spontaneously evolve the exchange of food. Some robots tend to live in the part of the environment which contains one type of food and other robots in the part of the environment that contains the other type of food, and then the robots meet together to exchange one type of food for the other type of food [22, 29].

Food is only one type of good, where a good is anything that human beings try with their behaviour to have. Human beings want to have many different goods because their goods are not only those that exist in nature but they produce always new goods by using the existing goods: clothes, homes, tools, cars, and many other things. The increase in the number of goods that human beings want to have has caused the invention of money. The invention of money can be simulated in the following way ([17], Chapter 11). We begin with a population of robots that want to have many different goods and, since a robot cannot produce all these goods, the robots must meet together to exchange their goods. But when two robots meet together to exchange their goods, one or both robots may not need the particular goods that the other robot has and, therefore, the exchange cannot take place. To solve this problem, the robots spontaneously invent money. At the end of the simulation, we find that one particular good is exchanged in all exchanges, and this good is money. All the robots want to have money because they can obtain all sort of goods from other robots in exchange for money. The exchange of goods has become buying and selling.

We conclude this section by mentioning two general characteristics of social behaviour which still need to be reproduced with human robots.

The social environment and the natural environment are very different environments and what human robots must do to obtain what they want from the two environments is very different [21]. To obtain what they want from the natural environment, they must simply act physically on the natural environment. To obtain what they want from another robot, they must change the other robot's brain. And if we ask what they must change in the other robot's brain, the answer is: its motivations. As we have seen in Section 4, what human robots do depends on their motivations and on the current strength of their motivations, and their behaviour is aimed at satisfying the motivation which currently has the greatest strength. Therefore, to obtain what it wants from another robot, a robot must change the current strength of the other robot's motivations. This is social behaviour: changing the motivations of others so that they do what one wants them to do. To change the motivations of other robots, a robot can send all sorts of sensory inputs to their other robots' brain. It can talk to them, it can modify its external physical appearance by dressing and by decorating its body, and it can express its emotions with its face, its voice, and its body.

The social environment has other characteristics which make it different from the natural environment. An important capacity of human beings is the capacity to predict the consequences of their behaviour and to decide to actually execute the behaviour only if they consider these consequences as good [6]. This capacity can be simulated with robots in the following way

[20]. The neural network of the robots has two additional set of internal neurons, the prediction neurons and the "good/bad" neurons. The predictions neurons are activated by the current sensory input and by a planned but still not physically executed behaviour in response to the current sensory input. The "good/bad" neurons are activated by the prediction neurons. When the robots' neural network receives a sensory input from the environment, it does not automatically responds to this input by executing some behaviour but it plans some behaviour, predicts its consequences, judges if these consequences are good or bad, and physically executes the behaviour only if they are good.

But there is an important difference between predicting the effects of one's behaviour on the natural environment and on the social environment. To predict what will happen in the natural environment, human beings must take into considerations only the sensory inputs which arrive to their sensory organs from the natural environment. To predict what another individual will do, they must take into consideration not only the sensory input which currently arrives to their sensory organs from the other individual but also the sensory input which currently arrives to the other individual's sensory organs and what are the other individual's motivations. And there is also another problem. Human beings are more different from one another than inanimate objects and this makes their behaviour more difficult to predict. Inanimate objects obey more or less the same laws and these are relatively simple laws which are not so difficult to discover. The behaviour of human beings does not only obey more complex laws but each individual is so different from all other individuals that his or her behaviour cannot be predicted by only using general laws.

6. What problems will human robots pose to human beings?

As we have already said, human robots will pose problems to human beings and these problems will be more serious than the problems posed to human beings by humanoid robots or, more generally, by robots. This last section is dedicated to a very brief discussion of these problems.

As we have already said, while humanoid robots must necessarily be physically realized to be useful to those who buy them, human robots may be useful to science even if they are only simulated in a computer. However, in the future, human robots will also be physically realized and, when this will happen, they will pose more serious problems to human beings than physically realized humanoid robots.

However, human robots can pose problems to human beings even if they are only simulated in a computer. Computers are interactive devices and, therefore, human beings will have the possibility to interact with the simulated human robots which are inside their computer. Today, many human beings—especially young human beings—spend much of their time by interacting with the digital environment rather than with the real environment. But when the digital environment will contain human robots, it is possible that a much greater number of human beings—of all ages—will prefer to live in a simulated social environment made of simulated human beings rather than in the real social environment made of real human beings.

And the simulated human robots may convince them to do what is not in their interests or they may want to damage them in other ways. One might object that human beings can always switch off the computer but the simulated human robots may convince them not to do so.

Can we control human robots so that they do not do what *we* don't want them to do? This is already a problem for robots as technologies but for robots as technologies the problem can be solved by emanating laws that prohibit the construction of certain types of robots. This cannot be done with robots as science. We can put limits to technology, but can we put limits to science?

Human robots—whether simulated in a computer or physically realized—may also pose embarrassing questions to human beings. If someone constructs a robot which is like my friend Gabriele, who is Gabriele, my friend or the robot? To really understand me, I must construct a robot which is like me? Who am I, I or a robot which is like me?

But the true danger for human beings of a robotic science of human beings is that it will let human beings know themselves as science knows nature. Human robots will not only demonstrate that human beings are only nature but they will project the cold light of science on everything that we are, do, think, and feel. According to the Greek philosopher Democritus, "truth lies in the abyss." Human robots will let human beings fall in the abyss.

7. Conclusion

Unlike humanoid robots that are practically useful physical artefacts which have some resemblances to human beings, human robots are computer simulations that must reproduce everything that human beings are and do and will make it possible for science to finally understand human beings and their societies as it understands nature. In this chapter, we have described a number of individual and social human phenomena which have already been reproduced by constructing human robots and societies of human robots and we have briefly discussed some of the very serious problems that human robots will pose to human beings.

Author details

Domenico Parisi

Address all correspondence to: domenico.parisi@istc.cnr.it

Institute of Cognitive Sciences and Technologies, National Research Council, Rome, Italy

References

[1] Nolfi S, Elman JL, Parisi D. Learning and evolution in neural networks. Adaptive Behaviour. 1994;**3**:5-28. DOI: 10.1177/105971239400300102

[2] Elman JL, Bates EA, Johnson MA, Karmiloff-Smith A, Parisi D, Plunkett K. Rethinking Innateness. A Connectionist Perspective on Development. Cambridge, MA: MIT Press; 1996

[3] Heinrich J, Heine SJ, Norenzayan A. The weirdest people in the world? Brain and Behavioural Science. 2010;**33**:61-135

[4] Duchon AP, Kaebling LP, Warren WH. Ecological robotics. Adaptive Behaviour. 1998;**6**: 473-507. DOI: 1177/105971239800600306

[5] Mirolli M, Parisi D. How can we explain the emergence of a language which benefits the hearer and not the speaker? Connection Science. 2005;**17**:307-324. DOI: 10.1080/0954 0090500177539

[6] Cecconi F, Parisi D. Neural networks with motivational units. In: From Animals to Animats 2. Cambridge, MA: MIT Press; 1992. pp. 167-181

[7] Arbib MA and Fellows JM. Emotions. From brain to robots. Trends in Cognitive Science. 2004;**6**:554-561. DOI: 10.1016/j.tics.2004.10.004

[8] Parisi D, Petrosino G. Robots that *have* emotions. Adaptive Behaviour. 2010;**18**:453-469. DOI: 10.1177/1059712310388528

[9] Denaro D, Parisi D. Cultural evolution in a population of neural networks. In: Marinaro M, Tagliaferri R, editors, Neural Nets. London: Springer; 1996

[10] Ugolini M, Parisi D. Simulating the evolution of artefacts. In Floreano F, Nicoud J-D, Mondada F, editors. Advances in Artificial Life. London: Springer; 1999. pp. 489-498. DOI:10.1007/3-540-48304-7_67

[11] Acerbi A, Parisi, D. Cultural transmission between and within generations. Journal of Artificial Societies and Social Simulation. 2006;**9**(1)

[12] Parisi D, Nolfi S. Sociality in embodied social agents. In Run R, editor. Cognition and Multi-Agent Interactions. Cambridge: Cambridge University Press; 2006

[13] Da Rold E, Petrosino G, Parisi D. Male and female robots. Adaptive Behaviour. 2011;**2**: 5-407. DOI: 1177/1059712311417737

[14] Cecconi F, Denaro D, Parisi D. Social aggregation in evolving neural networks. In: Castelfranchi C, Werner E, editors. Artificial Social Systems. London: Springer; 1994. pp. 41-54. DOI: 10.1007/3-540.58266-5_3

[15] Baldassarre G, Parisi D, Nolfi S. Distributed coordination of simulated robots based on self-organization. Artificial Life. 2006;**12**:298-311. DOI: 10.1162/artl.2006.12.3.289

[16] Baldassarre G, Nolfi S, Parisi D. Evolving mobile robots able to display collective behaviour. Artificial Life. 2003;**9**:255-267. DOI: 10.1162/106454603322392460

[17] Parisi D. Future Robots. Towards a Robotic Science of Human Beings. Amsterdam: John Benjamins; 2014

[18] Cangelosi D, Parisi D. The emergence of a 'language' in an evolving population of neural networks. Connection Science. 1998;**10**:83-97. DOI: 10.1080/0954000998116512

[19] Cangelosi A, Parisi D. The processing the verbs and nouns in neural networks: Insights from synthetic bran imaging. Brain and Language. 2004;**2**:401-408. DOI: 10/1016/S0093-934X(03)00353-3

[20] Cecconi F, Parisi D. Learning to predict the consequences of actions. In: Eckmiller R, Hartmann G, Hauske G, editors. Parallel Processing in Neural Systems and Computers. Amsterdam: Elsevier; 1990. pp. 237-240

[21] Cecconi F, Parisi D. Individual versus social survival strategies. Journal of Artificial Societies and Social Simulation. 1998;**1**(2)

[22] Delre SA, Parisi D. Information and cooperation in a simulated labour market. A computational model of the evolution of workers and firms. In: Salzano M, Colander D, editors. Complexity Hints for Economic Policy. London: Springer; 2007

[23] Gigliotta O, Miglino O, Parisi D. Groups of agents with a leader. Journal of Artificial Societies and Social Simulation. 2007;**10**:1-10

[24] Mirolli M, Parisi D. Language, altruism, and docility: How cultural learning can favour language evolution. In: Pollock JB, Bedau M, Husbands P, Ikegami T, Watson RA, editors, Artificial Life. Vol. 9. Cambridge, MA: MIT Press; 2004. pp. 182-187

[25] 25. Mirolli M, Parisi D. Talking to oneself as a selective pressure for the emergence of language. In: Cangelosi A, Smith ADM, Smith K, editors. Proceedings of the 6th International Conference on the Evolution of Language. Singapore: World Scientific; 2006. pp. 182-187

[26] Mirolli M, Cecconi D, Parisi D. A neural network model for explaining the asymmetries between linguistic production and linguistic comprehension. In: Vosniadou S, Kaiser D, Protopapas A, editors. Proceedings of the European Cognitive Science Conference Hillsdale, NJ: Erlbaum; 2007. pp. 670-675

[27] Parisi D, Cecconi F, Nolfi S. Econets: Neural networks that learn in an environment. Network. 1990;**1**:149-168. DOI: 10.1088/0954-898X/1/2003

[28] Parisi D. Cultural evolution in neural networks. IEEExpert. 1997;**12**:9-11. DOI:10.1109/64.608170

[29] Parisi D. What to do with a surplus. In: Conte R, Hegselman R, Terna P, editors. Simulating Social Phenomena. London: Springer; 1997. pp. 133-151. DOI: 10.1007./978-3-662-03366-1_10

[30] Parisi D. An artificial life approach to language. Brain and Language. 1997;**59**:121-146. DOI: 10.1006/brln.1997-1815

[31] Parisi D. Internal robotics. Connection Science. 2004;**16**:325-338. DOI: 10.1080/0954009412331314768. (328-354)

[32] Ziemke T. The role of emotions in biological and robotic autonomy. Biosystems. 2008;**91**:401-408. DOI: 10.1016/.biosystems.2007.15.015